Aktuelle Forschung Medizintechnik – Latest Research in Medical Engineering

Editor-in-Chief:
Th. M. Buzug, Lübeck, Deutschland

W0227602

Among future technologies with high innovation potential, medical engineering counts among those with above-average growth rates and is considered crises-proof. Computerization, miniaturization, and molecularization are essential trends in medical engineering. Computerization is the basis for medical imaging, image processing, and image-guided methods in surgery. Miniaturization plays an important role in the field of intelligent implants, minimally invasive surgery as well as in the development of new nanostructured materials in medicine. Molecularization is both a crucial element in the field of regenerative medicine and the so called molecular imaging. Cross-sectional technologies like nano- and microsystems technology as well as optical technologies and softwaresystems are, therefore, of high relevance.

This series for outstanding dissertations and habilitation treatises in the field of medical engineering covers clinical engineering and medical computer science as well as medical physics, biomedical engineering and medical engineering science.

Dierck Hillmann

Holoscopy

With a foreword by Gereon Hüttmann

 Springer Vieweg

Dierck Hillmann
Lübeck, Germany

Dissertation University of Lübeck, 2013

ISBN 978-3-658-06478-5   ISBN 978-3-658-06479-2 (eBook)
DOI 10.1007/978-3-658-06479-2

The Deutsche Nationalbibliothek lists this publication in the Deutsche Nationalbibliografie; detailed bibliographic data are available in the Internet at http://dnb.d-nb.de.

Library of Congress Control Number: 2014942839

Springer Vieweg
© Springer Fachmedien Wiesbaden 2014

Printed on acid-free paper

Springer Vieweg is a brand of Springer DE.
Springer DE is part of Springer Science+Business Media.
www.springer-vieweg.de

Preface by the series editor

The book Holoscopy by Dr. Dierck Hillmann is the 15th volume of the Springer-Vieweg series of excellent theses in medical engineering. The thesis of Dr. Hillmann has been selected by an editorial board of highly recognized scientists working in that field.

The Springer-Vieweg series aims to establish a forum for Monographs and Proceedings on Medical Engineering. The series publishes works that give insights into the novel developments in that field. Prospective authors may contact the Series Editor about future publications within the series at:

Prof. Dr. Thorsten M. Buzug
Series Editor Medical Engineering
Institute of Medical Engineering
University of Lübeck
Ratzeburger Allee 160
23562 Lübeck
Web: www.imt.uni-luebeck.de
Email: buzug@imt.uni-luebeck.de

Foreword

Optical coherence tomography (OCT) is one of the most successful applications of modern optics in clinical diagnostics. By using the spectral information of broad bandwidth radiation, scattering tissue is imaged with micrometer depth resolution. By OCT *in vivo* imaging of the layered structure of the retina was enabled for the first time and 20 years after its first publication OCT has become an essential technique for retinal imaging. Parallel detection of light backscattered from different depths and single photon sensitivity are key features for the success of OCT.

This book introduces lateral parallelization of OCT in order to addresses two current shortcomings of this technology. 1. Imaging speed is limited by the maximal permissible exposure of the focused beam which scans the tissue. 2. The depth of focus in which the image is recorded is coupled to the lateral resolution. The center of attention of this book is the introduction of holoscopy, the combination of OCT with digital holography, which overcomes both limitations. By experimentally implementing ideas of Emil Wolf from the end of the 60s and Adolf Fercher ten years later, the whole scattered light field from a tissue volume is holographically recorded at discrete wavelengths over a certain spectral range. Images of biological objects and first *in vivo* images of human tissue are presented for the first time.

The description of theory and background of holoscopy includes a rigorous and comprehensive depiction of OCT itself. In two further chapters new approaches for data reconstruction by Fourier transforms on non-equispaced data and for correction of group velocity dispersion or sample motion are discussed. These techniques are equally important for both, OCT and holoscopy.

This book links optical coherence tomography to digital holography and shows that both have more in common than up to now was anticipated. The unifying view of this book paves the way for improved biomedical imaging with scattered ballistic photons.

I can highly recommend this book to all scientists and graduate students interested in coherence based imaging of biological tissue.

Lübeck, April 2014   PD Dr. Gereon Hüttmann

Contents

List of figures   xv

List of tables   xix

Summary   xxi

Zusammenfassung   xxv

Symbols and notation   xxix

Abbreviations   xxxv

1. Introduction   1
   1.1. Optical imaging 2
   1.2. Optical coherence tomography 3
      1.2.1. Techniques 3
      1.2.2. Applications 4
   1.3. Advantages and limitations of OCT 4
   1.4. New approaches 5
   1.5. Holoscopy 6
   1.6. Structure of the thesis 6

2. Theory   9
   2.1. The Fourier transform 9
      2.1.1. The Hilbert transform and analytic signals 10
      2.1.2. The discrete Fourier transform (DFT) 11
         2.1.2.1. Symmetries of the DFT 12
         2.1.2.2. Relation to the analytic Fourier transform 13
         2.1.2.3. The fast Fourier transform (FFT) 14
      2.1.3. Fourier transforms on non-equispaced nodes 15

2.1.3.1. The non-equispaced DFT (NDFT) 15
2.1.3.2. Interpolation and FFT (iFFT) 16
2.1.3.3. The non-equispaced FFT (NFFT) 18
2.2. Sampling 19
2.2.1. Discretized signals 19
2.2.2. The sampling theorem 19
2.2.2.1. Aliasing 21
2.3. Scalar waves 21
2.3.1. Monochromatic waves 21
2.3.2. Diffraction and propagation 22
2.3.2.1. Diffraction and propagation in Fourier space . . . 22
2.3.2.2. The propagator 24
2.3.3. Broadband light 25
2.3.3.1. Dispersion 26
2.3.3.2. Group and phase velocity 27
2.3.4. Coherence 28
2.3.4.1. Temporal coherence 29
2.4. Coherent imaging 31
2.4.1. Image formation 31
2.4.2. Quadratic phase factors 33
2.4.3. Thin lenses 33
2.4.4. The Fresnel approximation 34
2.4.4.1. Conjugated planes 36
2.4.4.2. Relation to the angular spectrum 36
2.4.5. Lateral resolution 37
2.4.5.1. The point spread function 37
2.4.5.2. The Rayleigh criterion 37
2.4.5.3. The numerical aperture (NA) 38
2.4.5.4. Gaussian beams 39
2.5. Optical scattering theory 42
2.5.1. The Green's function 42
2.5.2. Born series 43
2.5.3. The Ewald's sphere 45
2.5.3.1. Diffraction grating 47
2.6. Holography 48
2.6.1. Classical holography 48
2.6.2. Digital holography 50
2.7. Optical coherence tomography (OCT) 50
2.7.1. Time-domain optical coherence tomography (TD-OCT) . . . 50
2.7.2. Fourier-domain optical coherence tomography (FD-OCT) . 51

	2.7.2.1.	The axial point spread function	55
	2.7.2.2.	Dispersion in FD-OCT	55
	2.7.2.3.	Sensitivity advantage of FD-OCT	56
	2.7.2.4.	Signal roll-off in FD-OCT	56
2.7.3.	Direct and heterodyne detection		57
	2.7.3.1.	Direct detection	57
	2.7.3.2.	Heterodyne detection	57
2.7.4.	Scanning OCT		58
	2.7.4.1.	Confocal gating	58
2.7.5.	Full-field OCT		60

3. FD-OCT signal processing using the NFFT — **63**

3.1.	The chirped FD-OCT signal		63
3.2.	Calibration		64
	3.2.1.	Calibration in presence of GVD mismatch	65
3.3.	Materials and methods		68
	3.3.1.	The algorithms	68
	3.3.2.	Simulation	68
	3.3.3.	Measured data	69
	3.3.4.	Signal processing	70
3.4.	Results and discussion		70
	3.4.1.	Processing speed	70
	3.4.2.	Image quality with simulated data	71
	3.4.3.	Image quality with measured data	73

4. Motion and dispersion correction in FD-OCT — **83**

4.1.	Introduction		83
	4.1.1.	Doppler effect on axial motion in swept-source OCT	84
	4.1.2.	Dispersion in FD-OCT	85
4.2.	Effect and correction of motion and GVD mismatch		86
4.3.	Determination of the correcting phase function		87
	4.3.1.	Cross-correlation of sub-bandwidth reconstructions	87
4.4.	Materials and methods		91
	4.4.1.	Implementation	91
	4.4.2.	Full-field FD-OCT setup for axial motion experiments	93
	4.4.3.	FD-OCT setup for GVD mismatch experiments	94
4.5.	Results and discussion		94
	4.5.1.	Axial motion in full-field SS-OCT	94
	4.5.2.	GVD mismatch in FD-OCT	97

Contents

Theory of holoscopy105

5.3.1. The intensity distribution on the camera 105
5.3.2. The phase-corrected propagator, object and reference field . 109
5.3.3. Obtaining the phase-corrected object wave field 111
5.3.3.1. Reducing the DC signals 111
5.3.3.2. On-axis geometry 112
5.3.3.3. Off-axis geometry 113
5.4. Reconstruction 121
5.4.1. Reconstruction of one Rayleigh length 121
5.4.2. One-step reconstruction 122
5.4.2.1. In free space 122
5.4.2.2. In a medium 123
5.4.2.3. With numerical magnification 124
5.4.2.4. The complete reconstruction integral 126
5.4.3. Resolution 129
5.4.3.1. Axial resolution 129
5.4.3.2. Lateral resolution 129
5.4.4. The Ewald's sphere in holoscopy 130
5.4.4.1. Relation between scattering theory and one-step
holoscopy reconstruction 131
5.5. Setup considerations 132
5.5.1. Restrictions on the lens-less setup by the Nyquist criterion . 132
5.5.2. Considerations with imaging optics 137
5.5.2.1. Reconstruction with a simulated lens 137
5.5.2.2. Lens-less reconstruction 139
5.6. Materials and methods 142
5.6.1. Implementation 142
5.6.1.1. Numerical complexity and execution speed 142
5.6.2. Calibration 144
5.6.3. Setup 145
5.6.3.1. Lens-less holoscopy 145
5.6.3.2. High-resolution holoscopy 146
5.7. Results 146
5.7.1. Lens-less holoscopy 146
5.7.2. High-resolution holoscopy 148
5.7.3. Artifacts in holoscopic imaging 148
5.7.3.1. Autocorrelation artifacts 148

5.7.3.2. Multiple scattering 150
5.7.3.3. Horizontal lines 152

6. Conclusion   **153**
6.1. FD-OCT signal processing using the NFFT 153
6.2. Motion and dispersion 154
6.3. Holoscopy 155
6.4. Outlook 155

A. Mathematical supplements   **157**
A.1. Definitions 157
A.2. Useful identities 159
A.3. The Fourier transform 159
     A.3.1. Properties of the Fourier transform 159
     A.3.2. Important Fourier transforms 161
     A.3.3. The uncertainty relation of Fourier analysis 162
A.4. Complex analysis 164
     A.4.1. Residual theorem 165
     A.4.2. Jordan's lemma 166

B. Resolution in signal processing and optics   **167**
B.1. Apodization windows 167
     B.1.1. Rectangular window 167
     B.1.2. Hann window 168
     B.1.3. Ideal Gaussian window and OCT resolution 169
     B.1.4. Truncated Gaussian window 170
     B.1.5. Comparison of window functions 171
B.2. Lateral resolution of an imaging system 171
     B.2.1. Circular aperture 171
     B.2.2. Rectangular aperture 173
     B.2.3. Gaussian beams 175

C. Discretization of the reconstruction   **179**
C.1. General considerations 179
     C.1.1. Sampling theorem for discretization of analytical functions . 179
     C.1.2. Bandwidth of a convolution 181
     C.1.3. Zero padding 181
C.2. The acquired image and wave fields 182
     C.2.1. Obtaining the object wave-field 182
     C.2.2. Time-frequency filter 183

C.2.3. Removing the reference wave field 183
C.2.4. Spatial filter 184
C.3. The angular spectrum 186
C.3.1. Propagation 187
C.4. Reconstruction 187
C.4.1. Single reconstruction 187
C.4.2. Complete one-step reconstruction 188

Bibliography   **191**

Publications   **203**

Danksagung   **205**

List of figures

2.1.1. Interpolated FFT and non-equispaced FFT interpolation by a convolution. 18
2.3.1. Schematic drawing of a Michelson interferometer. 30
2.4.1. The numerical aperture (NA) of an imaging system. 39
2.4.2. Schematic drawing of a Gaussian beam. 41
2.5.1. The Ewald's sphere in optical scattering theory. 47
2.7.1. Simulated TD-OCT interference signal of a mirror. 52
2.7.2. Setup of a fiber based time-domain OCT system. 52
2.7.3. Setup of a spectrometer-based FD-OCT system. 53
2.7.4. Simulated FD-OCT spectrum and resulting A-scan of a mirror. 55
2.7.5. Principle of confocal gating. 59
2.7.6. Full-field time-domain OCT setup using a Linnik interferometer. . . . 59
2.7.7. Full-field swept-source OCT setup. 61

3.2.1. Calibration of chirp and dispersion for OCT with a single measurement from a cover slip. 67
3.4.1. Obtained A-scan rate for various algorithms for OCT signal processing. 71
3.4.2. Simulated chirped FD-OCT signals and their resampled FFTs. 72
3.4.3. Influence of imaging depth and SNR on imaging artifacts of A-scans computed from chirped FD-OCT signals. 74
3.4.4. Comparison of algorithms for resampling and FFT for a high-SNR OCT system. 76
3.4.5. Comparison of resampling and FFT algorithms for a high-SNR OCT image of an IR viewing card. 77
3.4.6. Comparison of algorithms for resampling and FFT for a low-SNR OCT system. 79
3.4.7. Comparison of algorithms for resampling and FFT for a low-SNR OCT system using a linear-k-spectrometer. 81

3.4.8. Comparison of resampling and FFT algorithms for a low-SNR OCT image of a contact lens.   82

4.1.1. Examples of motion artifacts in full-field SS-OCT.   84
4.1.2. Example of the effect of dispersion mismatch between reference and sample arm in FD-OCT.   84
4.3.1. Demonstration of the effect of dispersion on the FD-OCT signal and its short-time Fourier transform.   88
4.3.2. Diagram demonstrating cross-correlation of sub-bandwidth reconstructions.   90
4.4.1. Schematic representation of lateral filtering for OCT artifact removal.   93
4.5.1. Motion correction of a motion-blurred full-field SS-OCT measurement.   95
4.5.2. Motion correction of a strongly motion-blurred full-field SS-OCT measurement.   96
4.5.3. FD-OCT data of a finger tip with GVD mismatch and its correction using CCSBR.   98
4.5.4. Correction of GVD mismatch by CCSBR in retinal high-resolution FD-OCT imaging.   100
4.5.5. Axial PSFs after correction for GVD mismatch in FD-OCT using calibration and CCSBR.   101

5.1.1. Efficient use of photons: comparison between confocal systems and holoscopy.   104
5.2.1. Schematic setup drawings for holoscopy.   106
5.3.1. Coordinate system used for derivations and computations of the wave fields in holoscopy.   109
5.3.2. Separation of different interference terms in off-axis holography and holosocpy.   115
5.3.3. Bandwidths of the different interference terms in off-axis holography and holoscopy.   118
5.3.4. Demonstration of spatial filtering and the resulting image shift for different wavelengths in off-axis holoscopy.   120
5.4.1. Focal shift in a medium with a refractive index.   124
5.4.2. Coordinate transformation required for one-step reconstruction in holoscopy.   127
5.4.3. NA of a lens-less holoscopy setup.   130
5.4.4. The Ewald's spheres in holoscopy.   131
5.5.1. Setup considerations for lens-less holoscopy, required for optimal sampling.   134

5.5.2. Maximum possible NA for a concrete lens-less holoscopy setup, as function of various setup parameters. 135

5.5.3. General restriction of the NA in lens-less holoscopy. 136

5.5.4. Geometrical considerations for holoscopy using a microscope objective. 138

5.5.5. Possible transformation of the setup geometry for high-resolution holoscopy. 140

5.5.6. DH reconstruction with angular spectrum approach and paraxial approximation. 141

5.6.1. Improvement of holoscopy processing speed by efficient one-step reconstruction. 144

5.6.2. Reconstructed aperture to calibrate setup distances. 145

5.7.1. Holoscopy measurement of an artificial point scattering sample demonstrating different reconstruction algorithms. 147

5.7.2. Tomographic en-face images of a bug, acquired by holoscopy. 149

5.7.3. Holoscopy measurement of a finger tip. 150

5.7.4. High-resolution holoscopy images of a grape showing different reconstruction algorithms. 151

5.7.5. Imaging artifacts in holoscopy. 152

B.1.1. Axial window functions and their Fourier transforms. 172

B.1.2. Axial point spread functions for various windows. 173

B.2.1. Apertures, two-dimensional apodization windows and their according lateral point spread functions. 174

C.2.1. Schematic showing the data layout when spatial filtering is applied. . 185

List of tables

4.5.1. Comparison of methods to obtain GVD mismatch in FD-OCT. 102

Summary

Tomographic imaging gives insight into the living human body. Among the numerous tomographic techniques, invented in the last century, optical methodologies take a special role. They provide high-resolution imaging, down to the order of a micron and are innocuous for the patient. Optical coherence tomography (OCT), invented in the 1990s, has found widespread adaption, especially in ophthalmology and its growth still continues. Research is ongoing and indicates its usefulness in many more medical and industrial disciplines.

The most important advance in OCT-technology was an increase in imaging speed by several orders of magnitude during the last decade by the introduction of Fourier-domain (FD) OCT, where the spectrally resolved interference signal of scattered light from a sample and reference light is measured. The aim of the work presented in this thesis was to develop technology to further increase imaging speed by efficient data evaluation and parallelization of OCT imaging.

The central part in the signal processing chain of modern FD-OCT is a discrete Fourier transform which usually needs to be performed on non-equispaced data nodes. In the first part of the thesis, three different algorithms to perform this task are compared and evaluated in terms of processing speed and resulting imaging quality: (1) the mathematical precise discrete Fourier transform on non-equispaced nodes (NDFT), (2) the linear interpolation followed by a fast Fourier transform (iFFT), and (3) the fast Fourier transform on non-equispaced nodes (NFFT) which comprises advanced interpolation and post-processing steps. The algorithms are compared with simulated data and data acquired by three different OCT devices: a low-speed system with a high signal-to-noise ratio (SNR), a high-speed low-SNR system and a high-speed low-SNR system using a linear-k-spectrometer, which reduces the non-linearity of the data nodes significantly. Imaging artifacts, side lobes of the signal peaks and a decreased SNR due to an increase in numerical noise were observed and quantified. A trade-off between image quality and processing speed was observed. The optimal processing thereby depends on the OCT device. The use of the linear-k-spectrometer increases imaging quality when using fast processing algorithms, but renders the OCT system more vulnerable to imaging artifacts created by the camera or by electronic noise.

Today, imaging speed of FD-OCT is limited by the scanning speed and maximum permissible exposure (MPE) of the tissue. Full-field FD-OCT circumvents these limitations. It acquires data in parallel by using an area camera, and spectral resolving of the interference measurement is achieved by using a swept-source laser. In this scenario, the fast and the slow axes of data acquisition are exchanged, making the acquisition much more vulnerable to motion of the specimen, resulting in imaging artifacts and image blurring. Because of the Doppler effect, this technology is especially vulnerable to axial motion. In the second part of this thesis, an algorithm is proposed to detect the motion of the specimen and to correct for the resulting imaging artifacts without requiring additional hardware. By using a short-time Fourier transform, sub-regions of the spectrum, in which motion of the specimen is limited, are windowed and Fourier transformed individually. The resulting reconstructed OCT images are cross-correlated to determine their apparent displacement. Using this data, a correction phase factor is computed, which is used to revert the effects of the axial motion. The algorithm is successfully demonstrated on motion blurred full-field FD-OCT data.

Axial motion of the specimen has the same effect on swept-source OCT data as group velocity dispersion (GVD) mismatch between reference and sample arm, which is a common problem in OCT measurements, that causes loss of signal and axial resolution. The developed algorithm allows to also determine and correct GVD mismatch. Individual correction of GVD mismatch, introduced by the human eye bulb, is demonstrated. For ultra-high-resolution OCT imaging of the retina an improved imaging quality and axial resolution was observed.

A fundamental limitation of FD-OCT is the limited depth of focus, which reduces lateral resolution and sensitivity outside of the focal range. When increasing the lateral resolution, the usable depth of field decreases with the square of the lateral resolution.

Digital holography (DH) is able to record and reconstruct images without limitation to a focal region, because it captures the complete information of the scattered field. Numerical propagation can bring all parts of the image to focus. In the third and final part of this thesis, the techniques of digital holography are combined with the optical sectioning of Fourier-domain OCT, resulting in a new imaging modality, "holoscopy". Tomographic *ex vivo* and *in vivo* images at low and moderate numerical aperture (NA) are shown with full lateral resolution and sensitivity over more than 30 Rayleigh lengths. The combination of digital holography with OCT results in an increased numerical complexity: acquired digital holograms for more than 1000 wavelengths were refocused numerically to every single focus within the sample, and for each focus a standard FD-OCT reconstruction followed. A single step reconstruction process via resampling in the Fourier space is proposed, which reduced processing time by orders of magnitude. This way, the time required for a reconstruction of a complete volume was decreased to less than a minute, independent of the resulting resolution.

Currently, holoscopy still has limitations. The available tunable lasers provide insufficient spectral tuning range, resulting in insufficient axial resolution to fully use the advantages of holoscopy, which become more apparent in high-resolution imaging. In addition, the available cameras are too slow or are too expensive for widespread clinical use. Most severely, the lack of a confocal detection scheme makes holoscopy vulnerable to incoherent background light and multiple scattered photons, resulting in a loss of imaging quality.

While holoscopy has some principle drawbacks compared to FD-OCT, the technology has potential. The additional increase in imaging speed and resolution, while using a simpler setup at the cost of more numerical post-processing, reminds of the change from time-domain OCT to FD-OCT. Finally, holoscopy allows further improvements to imaging, for example, aberration correction can be applied by only using suitable algorithms – without requiring expensive hardware.

Zusammenfassung

Tomographische Bildgebung ermöglicht Einsichten in den lebenden menschlichen Körper. Unter den zahlreichen tomographischen Techniken, die im Laufe des letzten Jahrhunderts entwickelt wurden, spielen optische Methoden eine besondere Rolle. Sie ermöglichen höchstauflösende Bildgebung von Strukturen im Mikrometerbereich und sind harmlos für den Patienten. Optische Kohärenztomographie (OCT, engl. optical coherence tomography), erfunden Anfang der neunziger Jahre, hat heutzutage zahlreiche Anwendungen, insbesondere im Bereich der Ophthalmologie. Aktuelle Forschung deutet darauf hin, dass OCT sich auch in zahlreichen weiteren medizinischen und industriellen Anwendungen als nützlich erweisen wird.

Der wichtigste Fortschritt der OCT-Technologie bislang, war eine Erhöhung der Bildgebungsgeschwindigkeit um einige Größenordnungen, die sich im Laufe des letzten Jahrzehnts mit der Entwicklung der Fourier-Domain (FD) OCT abzeichnete. Bei dieser Technik werden spektral aufgelöste Interferenzsignale aus einem Proben- und einem Referenzstrahl aufgenommen und zu Tomogrammen verrechnet. Das Ziel dieser Arbeit war es, die Bildgebungsgeschwindigkeit weiter zu erhöhen, indem eine effiziente Datenverarbeitung etabliert, und die Datenerfassung weitestgehend parallelisiert wurde.

Der zentrale Teil der FD-OCT-Datenverarbeitungskette ist eine diskrete Fourier Transformation (DFT), die im allgemeinen auf nichtlinear verteilten Stützstellen durchgeführt werden muss. Im ersten Teil der Arbeit, werden drei Algorithmen für diese Aufgabe verglichen, mit dem Ziel Datenverarbeitungsgeschwindigkeit und Bildqualität zu optimieren: (1) die mathematisch exakte Behandlung mittels einer diskreten Fourier Transformation auf nicht-äquidistanten Stützstellen (NDFT, engl. non-equispaced discrete Fourier transform), (2) die lineare Interpolation, gefolgt von einer schnellen Fourier Transformation (iFFT, engl. interpolated fast Fourier transform) und (3) die schnelle Fourier Transformation auf nicht-äquidistanten Stützstellen (NFFT, engl. non-equispaced fast Fourier transform), die auf einer erweiterten Interpolation, gefolgt von einer gewöhnlichen schnellen Fourier Transformation (FFT, engl. fast Fourier transform) und einer Nachverarbeitung der Daten beruht. Die Algorithmen wurden mit simulierten und gemessenen Daten ausgewertet und verglichen. Die Messungen wurden mit drei verschiedenen OCT-Systemen durchgeführt: Ein vergleichsweise langsames System mit einem großen Signal-Rausch-Verhältnis (SNR, engl. signal-to-noise ratio), ein vergleichsweise schnelles System mit einem niedrigen SNR, und ein schnelles System mit einem niedrigen SNR und einem speziellen Linear-k-Spektrometer, welches die nichtlineare Verteilung der Stützstellen für die FFT signifikant verringert. Bildartefakte, Seitenbänder der Signale und ein verringertes SNR durch ein erhöhtes numerisches Rauschen wurden beobachtet und quantifiziert. Der optimale Algorithmus hängt dabei vom verwendeten OCT-System ab, und stellt einen Kompromiss aus Bildqualität und Prozessierungsgeschwindigkeit dar. Das Linear-k-Spektrometers erhöht die Bildqualität, wenn schnelle Algorithmen benutzt werden, allerdings ist ein solches System anfälliger für Bildartefakte durch die Kamera und durch elektronisches Rauschen.

Heutzutage ist die Bildgebungsgeschwindigkeit von FD-OCT hauptsächlich durch die Geschwindigkeit der lateralen Abtastung (Scannen) und die maximal zulässige Bestrahlung (MZB) limitiert. Full-Field FD-OCT umgeht diese Limitierungen. Hierbei werden die Daten parallel mit einer Flächenkamera aufgenommen, die spektrale Auflösung der Interferenzmessung wird mit einem durchstimmbaren Laser erreicht. Allerdings sind in diesem Fall die schnelle und die langsame Achse der Datenerfassung vertauscht, und die Bildgebung ist spürbar empfindlicher für Bewegungen der Probe. Dies führt zu Bildartefakten und verwaschenen Bildern. Bedingt durch den Dopplereffekt reagiert diese Technologie dabei besonders empfindlich auf axiale Bewegungen. Im zweiten Teil der Arbeit wird daher ein Algorithmus vorgestellt, der diese Bewegung detektiert und korrigiert – ohne Verwendung weiterer Hardware. Mittels einer Kurzzeit-Fourier-Transformation (STFT, engl. short-time Fourier transform) werden zunächst Bereiche des Spektrums gefenstert. Für diese einzelnen Bereiche ist die Bewegung der Probe vernachlässigbar und das spektrale Signal kann daher ohne Bewegungsartefakte Fourier-transformiert werden. Die so erhalten Kurzzeit-Bilder der Messung werden miteinander kreuzkorreliert um ihre Verschiebung zu verschiedenen Zeiten zu bestimmen. Aus diesen Daten werden Phasenfaktoren berechnet, mit denen die Bewegungsartefakte entfernt werden. Der Algorithmus wird erfolgreich an Full-Field FD-OCT Daten demonstriert.

Axiale Bewegung der Probe hat auf das gemessene Swept-Source OCT-Signal denselben Effekt, wie Unterschiede der Gruppengeschwindigkeitsdispersion (GVD, engl.: group velocity dispersion) zwischen Referenz- und Probenarm. Diese stellen ein häufiges Problem in der OCT-Bildgebung dar und verschlechtern die axiale Auflösung. Folglich kann der entwickelte Algorithmus ebenfalls verwendet werden, um GVD-Unterschiede zu messen und zu korrigieren. Eine individuelle Korrektur des GVD-Unterschiedes, verursacht durch die Länge des Augapfels, wird anhand von hochauflösenden OCT-Aufnahmen der Retina demonstriert; die axiale Auflösung und die Bildqualität wurden verbessert.

Eine fundamentale Limitierung von FD-OCT ist die begrenzte Fokustiefe: Laterale Auflösung und Sensitivität verringern sich außerhalb des Fokusbereichs. Dabei reduziert sich die Tiefenschärfe quadratisch mit der lateralen Auflösung.

Mittels Digitaler Holographie (DH) kann man Bilder ohne Limitierung der Tiefenschärfe aufnehmen und rekonstruieren, da die gesamte Information des gestreuten Lichtfeldes aufgezeichnet wird. Durch numerisches Propagieren können alle Teile des Bildes nachträglich fokussiert werden. Im dritten und letzten Teil der Arbeit werden die Techniken der DH mit den Techniken der FD-OCT zu einer neuen Bildgebungsmodalität, "Holoskopie", kombiniert. Tomographische *ex vivo* und *in vivo* Aufnahmen bei geringer und moderater numerischer Apertur (NA) werden gezeigt. Die volle laterale Auflösung und Sensitivität wird dabei über mehr als 30 Rayleighlängen aufrechterhalten. Allerdings macht die Kombination von DH mit FD-OCT die numerische Rekonstruktion deutlich aufwändiger. Über 1000 Hologramme, aufgezeichnet bei verschiedenen Wellenlängen, mussten in jede Fokusebene der Probe numerisch propagiert werden, und für jede dieser Fokusebenen musste anschließend eine normale FD-OCT-Rekonstruktion erfolgen. Zur Umgehung dieser langen Rekonstruktionszeiten wird daher ein Einschritt-Rekonstruktionsverfahren mittels Interpolation im Fourier-Raum vorgestellt, welches die Rekonstruktionszeit um Größenordnungen reduziert. Mit diesem benötigt die Rekonstruktion weniger als eine Minute, unabhängig von der verwendeten Auflösung.

Zum jetzigen Zeitpunkt sind die Möglichkeiten der Holoskopie begrenzt. Die aktuell erhältlichen Laser haben keinen ausreichenden Durchstimmbereich und damit ermöglichen sie keine ausreichende axiale Auflösung, um die Vorteile der Holoskopie voll auszunutzen. Zudem sind die erhältlichen Kameras zu langsam oder zu teuer, um eine breite, klinische Anwendung zu ermöglichen. Am dramatischsten erweist sich jedoch das Fehlen der konfokalen Detektion. Dies führt dazu, dass die Bildgebung mittels Holoskopie empfindlich auf inkohärentes Hintergrundlicht und mehrfachgestreute Photon ist, was zu einem Verlust an Bildqualität führt.

Doch trotz einiger prinzipieller Nachteile der Holoskopie gegenüber FD-OCT, hat die Technologie Potential. Die Steigerung der Bildgebungsgeschwindigkeit und Auflösung, bei einem vergleichsweise einfachen Aufbau und auf Kosten eines erhöhten numerischen Rechenaufwands, erinnert an den technologischen Fortschritt von Time-Domain OCT zu FD-OCT. Letztendlich lassen sich mit Holoskopie weitere Technologien leicht verbinden. Beispielsweise kann Aberrationskorrektur (adaptive Optik) nur durch die Verwendung geeigneter Algorithmen angewandt werden – ohne das hierfür kostspielige Hardware benötigt wird.

Symbols and notation

Remarks on notation used throughout this thesis

- Vectors will be denoted in bold italic face and its x, y, and z-components with a subscript indicating the component. For example, the components of the vector $\boldsymbol{k}$ are k_x, k_y, and k_z. An exception is the vector denoting a position, $\boldsymbol{x}$, where the components will be denoted x, y, and z.

- Scalar fields will be called either with all components separately or as a collective vector parameter, i.e. $U(x, y, z)$ or $U(\boldsymbol{x})$.

- Operators will be denoted in scripted typeface and the function it is acting on will be set in squared brackets, for example $\mathscr{F}[f(x)]$.

- Inverse operators, functions and matrices will have the superscript -1, as for example $\mathscr{F}^{-1}[\cdot]$ for the inverse Fourier transform.

- The Fourier transform as a special operator will be denoted $\mathscr{F}[\cdot]$. In most cases it should be clear with respect to which axis the Fourier transform is taken. In cases where this is not entirely clear an additional subscript is used, for example $\mathscr{F}_{xy}[\cdot]$ indicates a two-dimensional Fourier transform with respect to the x and y-axis. Additionally, it should be clear from the text what the Fourier conjugated variable corresponding to a certain variable is. In most cases the Fourier conjugated variable corresponding to x, y, and z are k_x, k_y, and k_z, respectively. In case of a one-dimensional Fourier transform, the conjugated variable to x or z are sometimes both denoted k, if no confusion can arise.

- The Fourier transform of a function f is denoted with its corresponding tilded symbol, i.e. $\tilde{f}$. For multi-dimensional scenarios confusion might arise. In this case the function parameters indicate the axis in which the Fourier transform was taken, i.e. $\tilde{U}(k_x, k_y, z)$ denotes the two-dimensional Fourier transform of $U(x) = U(x,y,z)$ with respect to the x and y-axis. $\tilde{U}(x,y,k_z)$ denotes the one-dimensional Fourier transform with respect to the z-axis, etc. In case the notation is not entirely obvious, the axes that have been Fourier transformed are indicated by additional subscripts, e.g. $\tilde{U}_{xy}(k_x, k_y, z)$ is the two-dimensional Fourier transform of $U(x)$ with respect to the x and y-axis.

o All integral signs $\int$ represent improper integrals from $-\infty$ to $+\infty$, unless stated otherwise. In case an anti-derivative is meant, this is either obvious or stated in the text.

o Integrals are denoted in operator notation, i.e. $\int dx\, f(x)$ instead of $\int f(x)\, dx$. Multi-dimensional integrals are either denoted $\int d^3x \ldots$ for integration over the complete vector space or by each component on its own if confusion might arise, i.e. $\int dx \int dy \int dz \ldots$.

o It is not distinguished between matrices and its elements, e.g. $M_{\ell n}$ is used to identify the matrix M in the text.

o For discrete fields, describing images or their Fourier transforms, continuous variables such as x, y, z, k_x, k_y, and k_z need to be replaced with indices. In general ℓ, m, and n index the x, y, and z-coordinates, respectively. The indices decorated with a prime, ℓ', m', and n' index the corresponding Fourier conjugated variables k_x, k_y, and k_z. In some cases n' is used to index the absolute wavenumber k. Interpolated variables corresponding to a resampling of the n' coordinate are indexed by ν'.

Symbols

Below is a list of the most frequent symbols used in the thesis. Other symbols are explained in the text, as they occur.

Symbol	Meaning
$x = (x,y,z)$	spatial vector, Fourier conjugated to k
$k = (k_x, k_y, k_z)$	frequency vector, Fourier conjugated to x
$k, \|k\|$	absolute wavenumber
k_i, k_f	initial and final wavenumber of a sweep
λ	wavelength
λ_i, λ_f	initial and final wavelength of a sweep

Symbol	Meaning
t	time/parameter for which data was acquired, Fourier conjugated to ω
ω	circular frequency or Fourier conjugate to parameter t
ℓ, m	indices of a discretized wave field corresponding to x and y coordinates
n	index in a discretized wave field corresponding to z coordinate
ℓ', m'	indices of discretized angular-spectrum of a wave field corresponding to k_x and k_y coordinates
n'	indices of a discretized Fourier transformed wave field corresponding to k or k_z coordinate
N	number of images acquired
$I(x,y,t)$, $I(t)$	intensity value as acquired by a camera at pixel (x,y) or by a photodiode at time/parameter t
$U(z)$, $U(x,y)$, $U(x,y,t)$, $U(x,y,k)$	general wave field, at coordinates x, y, z, time t, and wavenumber k.
$\tilde{U}(k)$	Fourier transform of the wave field $U(x)$
$\tilde{U}_{xy}(k_x,k_y,z)$	Fourier transform of the wave field $U(x)$ with respect to the x and y-axis.
$R(x,y)$, $R(x,y,t)$	reference wave field in a given plane, e.g. the camera plane; acquired at time t
$R_0(x,y,k)$	phase-corrected reference wave field in a given plane, e.g. the camera plane at a certain wavenumber k
$O(x,y)$, $O(x,y,t)$	object wave field in a given plane, e.g. the camera plane; acquired at time t
$O_0(x,y,k)$	phase-corrected object wave field in a given plane, e.g. the camera plane at a certain wavenumber k
$i_{\ell mn}$	discretized form of the intensity field $I(x)$
$u_{\ell mn}$, u_n	discretized form of the wave field $U(x)/U(z)$

Symbol	Meaning
$\tilde{u}_{\ell'm'n'}$, $\tilde{u}_{n'}$	discretized form of the Fourier transforms of the wave field U, i. e. $\tilde{U}(k)$, $U(k)$
$o_{\ell mn}$	discretized form of the wave field $O_0(x)$
$r_{\ell mn}$	discretized form of the wave field $R_0(x)$
$\kappa_\zeta(k_x, k_y, k)$	Fourier space interpolation function in holoscopy
ζ	parameter describing proportionality between focus depth and optical path length in holoscopy
$A_C(k)$	normalized complex amplitude spectrum
A_R, A_O, A_I	amplitude of reference, object and incident wave field
$S(k)$	normalized intensity spectrum
c	speed of light
n, $n(\omega)$, $n(x)$	refractive index (as a function of ω/x)
$\eta(x)$	scattering potential
$\eta_{z_p}(x)$	reconstructed scattering potential for the focus depth
$\mathscr{F}[\cdot]$	Fourier transform
$\mathscr{F}^{-1}[\cdot]$	inverse Fourier transform
$\mathscr{F}_z[\cdot]$, $\mathscr{F}_{xy}[\cdot]$	Fourier transform with respect to the z, or the x and y-axis
$\mathscr{H}[\cdot]$	Hilbert transform
$\mathscr{P}_{k,z}[\cdot]$	propagator for the distance z and wavenumber k
$\mathscr{P}^0_{k,z}[\cdot]$	phase-corrected propagator for the distance z and wavenumber k
$P_{k,z}(k_x, k_y)$	propagation phase-function in Fourier space for the distance z and wavenumber k
$P^0_{k,z}(k_x, k_y)$	phase-corrected propagation phase-function in Fourier space for the distance z and wavenumber k
$Q_a(x, y)$	two-dimensional quadratic phase factor
$F_{n'n}$	Fourier matrix
$F^{-1}_{n'n}$	inverse Fourier matrix
$F_{\ell'\ell m'm}$	Fourier tensor
$\delta(x)$, $\delta(x)$	δ-distribution
$\Theta(x)$	step function

Symbol	Meaning
z^*	complex conjugated to z
$f * g$	convolution of f and g
$f \star g$	cross-correlation of f and g
$\partial_z f$	partial derivative of f with respect z
$\langle \cdots \rangle_t$	time-average
$\mathcal{O}(\cdots)$	time-complexity of an algorithm
$\mathbb{R}$	set of all real numbers

Abbreviations

Abbreviation	Meaning
CCSBR	cross-correlation of sub-bandwidth reconstructions
CPU	central processing unit
CT	computerized tomography
DC	direct current
DFT	discrete Fourier transform
DH	digital holography
DHM	digital holographic microscopy
ECBO	European Conference on Biomedical Optics
FD-OCT	Fourier-domain optical coherence tomography
FF	full-field
FF-FD-OCT	full-field Fourier-domain optical coherence tomography
FF-OCT	full-field optical coherence tomography
FF-SS-OCT	full-field swept-source optical coherence tomography
FF-TD-OCT	full-field time-domain optical coherence tomography
FFT	fast Fourier transform
FFTW	Fastest Fourier Transform in the West
FWC	full well capacity
FWHM	full-width at half maximum
GCC	GNU compiler collection
GNU	GNU's not Unix
GPU	graphics processing unit
GVD	group velocity dispersion

Abbreviation	Meaning
iFFT α	(linearly) interpolated fast Fourier transform with oversampling α
ISAM	interferometric synthetic aperture microscopy
ISO	International Organization for Standardization
MPE	maximum permissible exposure
MRI	magnet resonance imaging
MZB	maximal zulässige Bestrahlung
NA	numerical aperture
NDFT	non-equispaced discrete Fourier transform
NFFT α, m	non-equispaced fast Fourier transform with oversampling α and cut-off parameter m
NPL	National Physics Laboratory
OCM	optical coherence microscopy
OCT	optical coherence tomography
PET	positron emission tomography
PSF	point spread function
RAM	random access memory
SD-OCT	spectral-domain optical coherence tomography
SIMD	single instruction, multiple data
SLD	superluminescent diode
SNR	signal-to-noise ratio
SSE	streaming SIMD extension
SS-OCT	swept-source optical coherence tomography
STFT	short-time Fourier transform
TD-OCT	time-domain optical coherence tomography
US	United States
USAF	United States Air-Force

1 Introduction

Tomographic imaging has revolutionized medicine. It provides insight into the living human body and thereby allows diagnoses – without requiring direct access to the tissue and without harming the patient. Computerized tomography (CT) and magnet resonance imaging (MRI) allow visualization of large organs or even the whole body, albeit the former at the cost of radiation exposure. The ingenuity of both inventions were rewarded with the Nobel prize in Physiology or Medicine, Allan M. Cormack and Sir Godfrey N. Hounsfield received the award in 1979 "for the development of computer assisted tomography" [1], and Paul Lauterbur and Sir Peter Mansfield in 2003 "for their discoveries concerning magnetic resonance imaging" [2]. Although the resolution of these techniques is limited, in the clinical use to millimeters, they have numerous applications throughout medicine and sciences. The principle of tomographic imaging is also used in positron electron tomography (PET) and scintigraphy.

Ultrasonic imaging, first used in the 1950s, provides higher resolution of about a fraction of a millimeter, with a limited imaging depth of many centimeters. Compared to CT and MRI its application is cheap and it therefore found widespread use in many medical centers. Today, applications of ultrasonic imaging are found in prenatal diagnostics and cardiology using Doppler imaging. But for observing and diagnosing the smallest structures of the human physiology, its resolution does not suffice. The relatively large wavelength of ultrasound is the fundamental limitation.

1

1.1 Optical imaging

The wavelength of light is below a micrometer and thereby several orders of magnitude lower than the wavelength of ultrasound. Using visible or near infrared light, instead of ultrasound, therefore promises higher resolution.

Optical imaging utilizing microscopes has been used for centuries to image structures sized in the order of a micrometer. Classical wide-field microscopes illuminate the entire sample area with light and either observe the transmitted or the scattered and reflected light, magnified by suitable lens systems to make the structures in the order of the wavelength visible. This magnified image can be observed with the human eye or with a camera. Within the last century the techniques of optical microscopy have seen tremendous advances.

Optical sectioning techniques were introduced to reduce the influence and background light of out-of-focus layers of the specimen. One of the most important sectioning techniques is scanning confocal microscopy. Here, a single spot of the sample is illuminated with a highly focused beam and this exact focal spot is then imaged onto a pinhole; only the light backscattered from the sharply imaged focal spot passes this pinhole and its intensity is measured. The sample is then scanned by moving the respective focal spot over the sample.

In addition to reflected, scattered, and transmitted light, new contrast mechanisms have been found. Fluorescence imaging can highlight certain structures. The illuminating light is absorbed by certain molecules of the observed tissue and the afterwards emitted light has lower energy and is therefore easily distinguished from the elastically scattered light.

Nonlinear microscopy illuminates the sample with a high-intensity beam. The non-linear interaction between light and tissue, i.e. two or multiple photon absorption in the focal layer, where the intensity is maximal, achieves optical sectioning. The spontaneous emission of the excited states can be detected and is easily distinguished from the directly scattered photons due to its decreased wavelength. Nonlinear microscopy therefore provides both, sectioning and an additional contrast mechanism.

Quite a different technique of optical imaging is digital holography (DH). Here, the object is illuminated with coherent light, but the backscattered light is not imaged onto the camera using lenses and objectives, i.e. no sharp image is obtained. Instead this object light is superimposed with a reference light wave and the resulting interference pattern is recorded. This interference pattern encodes the amplitude and phase of the light wave that was scattered by the object. With the help of a computer and the knowledge of the reference wave field, the object light wave can be computed and numerically focused. Digital holography has also been demonstrated with microscopic resolutions, referred to as digital holographic microscopy (DHM).

1.2 Optical coherence tomography

Real optical tomographic imaging of three-dimensional scattering structures was achieved in the 1990s by optical coherence tomography (OCT). It provides a resolution of a few microns and allows *in vivo* imaging of the human physiology by detecting light, that is reflected or scattered by the observed tissue. Similar to ultrasound, OCT uses the propagation time of reflected light to measure the depth of tissue structures [3], but in contrast to ultrasound it images contact-free.

1.2.1 Techniques

OCT is an interferometric technique, and signals are acquired by measuring the interference of light scattered by a sample and light reflected by a reference mirror. Most commonly, OCT devices are distinguished in the way they achieve discrimination of depths, and in the way they acquire data for different lateral positions.

The first OCT measurements were based on time-domain (TD) OCT. Here a broadband light source with a short coherence length is used, restricting interference to the case that the scattered light from the sample and the light reflected by the reference mirror traveled the same optical path length when reaching the detector. The interference contrast for different reference path lengths gives the depth-resolved backscattering intensity of the sample. By moving the reference mirror, i.e. axial scanning, all sample layers are probed. Most commonly, time-domain OCT systems were combined with a confocal detection scheme that provides lateral discrimination. This way, samples are acquired three-dimensionally.

Fourier-domain (FD) OCT improves upon this principle and removes the axial scanning. The interference signal is acquired for only one reference path length, but spectrally resolved. The spectral interference pattern is Fourier transformed which yields the depth-resolved scattering intensity profile (Section 2.7.2). Two major techniques to acquire the spectrally resolved interference pattern have emerged: First, spectrometer-based OCT, also referred to as SpectralRadar or Spectral-domain (SD) OCT, uses a spectrometer and a broadband light source. Secondly, swept-source (SS) OCT acquires the spectrally resolved interference sequentially by using a wavelength-tunable laser. As time-domain OCT, FD-OCT systems are commonly combined with a confocal detection and lateral scanning.

Full-field OCT uses a wide-field illumination and detection, instead of lateral scanning. The sample is illuminated with an extended beam, e.g. collimated, and the interference signal is detected with an area camera. Full-field time-domain OCT is commonly used with very broadband spatially and temporally incoherent light sources in a Linnik interferometer; it allows for very high axial and lateral resolutions. Full-field swept-source OCT allows for a very high acquisition speed due to the parallelization in both, axial and lateral direction.

1.2.2 Applications

FD-OCT quickly gained widespread clinical success, especially because of its unique combination of imaging speed, resolution, and imaging depth. For ocular diagnostics, imaging cornea, lens, and retina [4–8] of the human eye, it has become a standard procedure. Ongoing research indicates that OCT will find further applications in non-opthalmic areas, such as dermatological [9, 10] and oncological diagnostics [11], as surveillance and guidance during surgery [12–14], in gastrointestinal diagnostics using endoscopes [15], or in cardiovascular imaging using catheters [16,17]. OCT also found its way into material sciences. For example, it is being used to analyze coatings [18], to analyze materials [19,20], distinguish foil layers [21], or as surveillance during manufacturing of solar cells [22].

Full-field time-domain optical coherence microscopy allows imaging of excised tissue with an imaging quality and resolution comparable to histological imaging [23,24], but *in vivo* application remains difficult with this technique, although first results have been demonstrated [25,26].

1.3 Advantages and limitations of optical coherence tomography

For high-speed imaging the efficient detection of scattered radiation is most important. It determines at a certain imaging quality how fast the imaging can be, as the maximum permissible exposure (MPE) limits the illumination intensity on the sample. Ultimately, it influences sensitivity and signal-to-noise ratio of OCT imaging. Two major properties of imaging allow to maximize photon detection:

First, imaging all scattering structures in parallel. Confocal microscopy and time-domain OCT reject photons whose scattering does not originate from a specific point in the specimen. Structures lying behind each others have to be measured sequentially, wasting scattered photons. Confocally scanning Fourier-domain OCT solved this problem partially by detecting all depths within the Rayleigh length in parallel (see Section 2.7.2). Increasing the lateral resolution reduces the Rayleigh length of the Gaussian scanning beam and thus the depth of focus decreases rapidly (see Section 2.4.5.4). The lateral resolution is only optimal within the Rayleigh length; multiple scanning in different depths is again necessary.

Secondly, heterodyne detection after interference with reference radiation allows quantum noise limited sensitivity even in the presence of detector noise or background radiation, as will be shown in Section 2.7.3. Techniques that detect backscattered light directly, such as confocal microscopy, suffer significantly from detection noise and stray light. Interferometric techniques such as OCT encode image information in the interference of the reference and sample light. If the reference light intensity is significantly larger than the sample intensity and also larger than any stray light, the sample light detection is shot noise limited. In this case the imaging quality in one sample layer is not influenced by light from other layers.

1.4 New approaches

Manifold techniques have been proposed to achieve tomographic optical imaging with a depth of focus spanning several Rayleigh lengths. The field of view of FD-OCT has been extended, for example by using multiple foci [27] or non-diffracting beams [28–32], i.e. Bessel illumination to achieve an increased depth of focus at high lateral resolution. Imaging quality was comparable to conventional FD-OCT in these approaches, but photon detection efficiency was not improved over sequential scanning of different depths. Numerical post-processing was applied, either by inverse scattering in interferometric synthetic aperture microscopy (ISAM, [33–35]), or by using algorithms of digital holography [36,37]. They achieved an increased depth of focus for confocally scanning FD-OCT, but all techniques that use scanning with either Gaussian or Bessel beams suffer from reduced sensitivity. In case of Gaussian beams, the detection efficiency drops rapidly for structures outside the focal layers. In case of Bessel beams, the detection efficiency is reduced in all depth layers, because large parts of the beam-energy are deposited in the side lobes of the beam.

To achieve a constantly high sensitivity over the full measurement depth, full-field approaches have to be applied: a collimated beam needs to be used for illumination, an area detector needs to collect all backscattered light, gating techniques need to be abandoned. Full-field Fourier-domain OCT was demonstrated *ex vivo* and *in vivo* by using high-speed imaging cameras and tunable lasers [38,39]. Still, lateral resolution was optimal only in the focal plane of the imaging optics.

For maximal imaging speed, photons need to be detected and pinpointed to their scattering origin with full lateral and axial resolution in the whole three-dimensional volume without loosing photons by gating techniques. Digital holography (DH) decouples the focus layer from the imaging optics, as it captures entire wave fields and enables numerical propagation and the introduction of numerical lenses (see Sections 2.3.2.2 and 2.6.2). But on its own, DH does not provide tomo-graphic images. Approaches using techniques of DH with multiple wavelengths to obtain tomographic images were shown, but so far they either showed a non-uniform resolution over depth [40–42] or suffered from inefficient reconstruction algorithms and reduced imaging quality [43–45].

1.5 Holoscopy

Holoscopy, the consequent combination of swept-source FD-OCT with DH by using a tunable laser in a holography-like setup, achieves uniform sensitivity and lateral resolution over a large measurement depth, even at high lateral resolution. Its imaging quality is comparable to scanning FD-OCT in low scattering tissue. As full-field FD-OCT, it achieves a volumetric acquisition of data without rejecting or gating photons by using an area camera. No scanning is required and no moving parts slow down the acquisition process. It is only limited by the camera acquisition rate and the sweep rate of the laser.

In holoscopy, an algorithm reconstructs the entire volume without limitations due to the limited depth of focus. It is similar to inverse scattering for full-field FD-OCT, which has so far only been demonstrated on simulated data [46, 47]. But in contrast to inverse scattering, the reconstruction in holoscopy models a forward physical process, the imaging of lenses and the propagation of light, instead of creating an abstract forward scattering model and inverting it mathematically. As OCT, holoscopy does not treat the data reconstruction as mathematical inverse problem and therefore does not require regularization techniques.

1.6 Structure of the thesis

This thesis consists of three major topics: the efficient processing of FD-OCT data on non-equispaced nodes, the compensation of dispersion and motion in Fourier-domain and swept-source OCT, respectively, and the development of holoscopy. Data processing and dispersion compensation are generally applicable to FD-OCT as demonstrated in Chapter 3 and 4, but are also useful for an efficient holoscopy reconstruction. The basic mathematical and physical concepts and tools that are required for this thesis, are laid out briefly in Chapter 2. These include the theoretical basics of Fourier transforms, sampling, waves, diffraction, coherent imaging, scattering, digital holography and OCT.

Chapter 3 analyzes a common problem in FD-OCT signal processing. The spectral OCT data is in general not acquired in the correct coordinate system to allow efficient and fast computation of the tomographic data by a simple Fourier transform. Algorithms to correct this are evaluated and performance in terms of resulting image quality and processing speed is compared. Reconstruction of holoscopic data face a similar problem. The algorithms were successfully applied to holoscopy.

Chapter 4 describes a common theory of imaging artifacts that are due to sample motion or dispersion mismatch in sample and reference arm in full-field and scanning OCT, respectively. Based on this theory, an algorithm to compensate for these effects is proposed and its effectiveness is demonstrated: motion artifacts for *in vivo* full-field FD-OCT images are corrected. Additionally, it is demonstrated that this algorithm can improve FD-OCT imaging quality in retinal imaging due to an individual correction of dispersion caused by the eye bulb.

Chapter 5 shows how to combine FD-OCT and digital holography to holoscopy and how to reconstruct the acquired data efficiently. Results with a simple lens-less setup and with a setup for high-resolution imaging are demonstrated.

Finally, Chapter 6 contains conclusions and a short discussion of possibilities, potential, and challenges of holoscopy.

2 Theory

The following Chapter is an overview of the most important physical and mathematical backgrounds that are required in this thesis to treat the topics of OCT, digital holography and holoscopy. It also conveys the formalisms that will be used in the following Chapters. More detailed treatments of the topics presented in this Chapter are found in [48–54].

2.1 The Fourier transform

The Fourier transform plays a very significant and important role in the complete thesis, therefore a summary of its definition and its most important properties is given below. For a more complete overview see e.g. [48,49,55].

A complex valued function U of a real position x can also be represented as a related function $\tilde{U}$ of the conjugated spatial frequency k, the latter being called the frequency domain or Fourier-domain representation. Both representations contain all information of the function, but it will be shown throughout this thesis, that physical laws and dependencies take a simpler form using the appropriate one. The Fourier transform $\mathscr{F}[\cdot]$ is the operator relating the functions U and $\tilde{U}$, defined by

$$\tilde{U}(k) = \mathscr{F}[U(x)] = \int \mathrm{d}x\, U(x) \mathrm{e}^{-ikx}. \tag{2.1.1}$$

This transform can be interpreted as a development into orthogonal harmonic functions, namely sines and cosines, and it is a linear operator. The inverse operator

$\mathscr{F}^{-1}[\cdot]$ takes the function $\tilde{U}$ back from frequency space to position space

$$U(x) = \mathscr{F}^{-1}\left[\tilde{U}(k)\right] = \frac{1}{2\pi}\int dk\, \tilde{U}(k)e^{+ikz}.$$

The concept of the Fourier transform and the frequency space representation can easily be generalized to three-dimensional physical fields U, that are a function of position $x = (x, y, z)$

$$\tilde{U}(k) = \mathscr{F}[U(x)] = \int d^3x\, U(x)e^{-ik\cdot x},$$

where $k = (k_x, k_y, k_z)$ describes the spatial frequencies. A vector dependency of a field can also be described in frequency space along one or two axes, while along the remaining axis it is described in position space, for example by $\tilde{U}_z(x, y, k_z)$ or $\tilde{U}_{xy}(k_x, k_y, z)$. The according generalizations to the Fourier transform operators are straight forward by sequential one-dimensional transforms for each axis.

A list of the most important properties of the Fourier transform along with a list of the transforms of important functions is found in the Appendix A.3.

2.1.1 The Hilbert transform and analytic signals

A complex valued function $U(z)$ is said to be an analytic signal, if its Fourier transform $\tilde{U}(k)$ fulfills the property

$$\tilde{U}(k) = 0, \quad \text{for all } k \leq 0.$$

Assuming a measurement only acquires the real part of a complex valued function $U(z)$, i.e. $\operatorname{Re} U(z)$, the function $U(z)$ can in general not be obtained from the measured data. It follows by using

$$\operatorname{Re} U(z) = \frac{1}{2}[U(z) + U^*(z)],$$

that

$$\begin{aligned}
\mathscr{F}[\operatorname{Re} U(z)] &= \frac{1}{2}[\mathscr{F}[U(z)] + \mathscr{F}[U^*(z)]] \\
&= \frac{1}{2}(\tilde{U}(k) + \tilde{U}(-k)),
\end{aligned}$$

i.e. the Fourier transform of the acquired data is symmetric with respect to the origin. If the signal $U(z)$ is known to be analytic, it can be obtained by filtering the negative frequency components and increasing the positive frequency components by a factor two:

$$\tilde{U}(k) = 2\Theta(k)\mathscr{F}[\text{Re }U(z)],$$

where $\Theta(k)$ is the step function (A.1.6). By using $2\Theta(k) = 1 + \text{sgn}(k)$, with $\text{sgn}(\cdot)$ being given by (A.1.7), and writing the multiplication as convolution of their respective position space functions (see convolution theorem (A.3.1)), it follows

$$\tilde{U}(k) = \mathscr{F}[\text{Re }U(z)] + \mathscr{F}\left[\mathscr{F}^{-1}[\text{sgn}(k)] * \text{Re }U(z)\right].$$

With $\mathscr{F}^{-1}[\text{sgn}(k)] = i/(\pi z)$, the analytic signal $U(z)$ can be written in position space as

$$
\begin{aligned}
U(z) &= \text{Re }U(z) + i\frac{1}{\pi}\left(\frac{1}{z} * \text{Re }U(z)\right) \\
&= \text{Re }U(z) + i\frac{1}{\pi}\int dz' \frac{1}{z-z'}\text{Re }U(z').
\end{aligned}
$$

The imaginary part motivates the definition of the Hilbert transform

$$\mathscr{H}[f(z)] = \frac{1}{\pi}\int dz' \frac{f(z')}{z-z'}.$$

The analytic signal of an acquired real signal can be obtained by setting the imaginary part of the signal equal to the Hilbert transform of the acquired real signal.

2.1.2 The discrete Fourier transform (DFT)

In numerical computations the integrals defining the Fourier transform can hardly be computed: signals are not acquired in the entire spatial or frequency domain and they are not obtained arbitrarily dense, i.e. its number of nodes is limited. A discretized version of the Fourier transform, known as the discrete Fourier transform (DFT), is used instead. It takes an N-dimensional complex vector u_n as input (position space vector) and maps it onto an N-dimensional complex vector $\tilde{u}_{n'}$ (frequency space vector). Being a linear transform it can be written as a matrix multiplication, which is defined by

$$\tilde{u}_{n'} = \sum_{n=0}^{N-1} F_{n'n}u_n, \quad \text{with } F_{n'n} = \exp\left(-i\frac{2\pi}{N}n'n\right), \tag{2.1.2}$$

where $F_{n'n}$ denotes the Fourier matrix. Its inverse is obtained by matrix inversion to

$$u_n = \sum_{n'=0}^{N-1} F_{n'n}^{-1} \tilde{u}_{n'}, \quad \text{with } F_{n'n}^{-1} = \frac{1}{N} \exp\left(+i\frac{2\pi}{N}n'n\right).$$

The generalization to a multi-dimensional discrete Fourier transforms is achieved by multilinear mappings using tensors. For example, the Fourier transform $\tilde{u}_{n'm'}$ of a two-dimensional field u_{nm} of size $N \times M$ is given by

$$\tilde{u}_{n'm'} = \sum_{n=0}^{N-1} \sum_{m=0}^{M-1} F_{n'nm'm} u_{mn},$$

$$\text{with } F_{n'nm'm} = F_{n'n}F_{m'm} = \exp\left(-i2\pi\left(\frac{n'n}{N} + \frac{m'm}{M}\right)\right), \quad (2.1.3)$$

with $F_{n'nm'm}$ being called the two-dimensional Fourier tensor, which is obtained as tensor product of the two one-dimensional Fourier matrices. Its effect is equivalent to transforming each axis individually.

2.1.2.1 Symmetries of the DFT

By extending the scope of the Fourier matrix indices to all integer numbers, symmetries arise:

$$F_{n'n} = F_{(n'+N)n} = F_{n'(n+N)}. \quad (2.1.4)$$

These symmetries often cause confusion in signal processing, as it allows for an alternative definition of the DFT:

$$\begin{aligned}
\tilde{u}_{n'} &= \sum_{n=0}^{N/2-1} F_{n'n}u_n + \sum_{n=N/2}^{N-1} F_{n'n}u_n \\
&= \sum_{n=0}^{N/2-1} F_{n'n}u_n + \sum_{n=N/2-N}^{N-1-N} F_{n'n}u_n \\
&= \sum_{n=-N/2}^{N/2-1} F_{n'n}u_n
\end{aligned}$$

With this definition, indices of the input and output vectors run from $-N/2$ to $N/2 - 1$. The negative indices are commonly represented as negative frequencies, especially when displaying data in frequency space. The DC ($n' = 0$) component

2.1. The Fourier transform

is centralized in this representation. For most programming languages, that only support indices running from 0 to $N-1$, frequencies need to be shifted.

2.1.2.2 Relation to the analytic Fourier transform

Let the signal $U(x)$ be known at equidistant points $x = n \cdot \Delta x$, meaning the specific values $u_n = U(n \cdot \Delta x)$ are given for $n = 0 \dots N-1$. Sampling the function $U(x)$ at these discrete points (see Section 2.2), represented by

$$U(x) = \sum_{n=0}^{N-1} u_n \delta(x - n \cdot \Delta x)$$

and inserting this into the one-dimensional Fourier transform (2.1.1) gives

$$
\begin{aligned}
\tilde{U}(k) &= \int \mathrm{d}z\, e^{-ikz} \sum_{n=0}^{N-1} u_n \delta(x - n \cdot \Delta x) \\
&= \sum_{n=0}^{N-1} u_n \int \mathrm{d}z\, e^{-ikx} \delta(x - n\Delta x) \\
&= \sum_{n=0}^{N-1} u_n e^{-ikn\Delta x}.
\end{aligned}
$$

Assuming the outgoing signal is equispaced in the Fourier-domain with spacing Δk, given by $\tilde{u}_{n'} = \tilde{U}(n' \cdot \Delta k)$, one obtains the DFT

$$\tilde{u}_{n'} = \sum_{n=0}^{N-1} e^{-in'n\Delta k\Delta x} u_n. \tag{2.1.5}$$

Comparing (2.1.5) with (2.1.2) gives an important relation between sampling distance of input and output data, when computing the DFT on sample data:

$$\Delta k = \frac{2\pi}{N\Delta x} \tag{2.1.6}$$

If $u_n = 0$ for $n \neq 0, \dots, N-1$, the DFT of u_n can be computed by the analytic Fourier transform $\tilde{U}(k)$ via

$$\tilde{u}_{n'} = \tilde{U}(n'\Delta k) = \tilde{U}\left(n'\frac{2\pi}{N\Delta x}\right) = \sum_{n=-\infty}^{+\infty} e^{-\frac{2\pi}{N}n'n} u_n = \sum_{n=0}^{N-1} e^{-\frac{2\pi}{N}n'n} u_n.$$

The proof follows from (A.3.8), the Fourier transform of the comb-distribution is again a comb-distribution.

2.1.2.3 The fast Fourier transform (FFT)

When computing the DFT of a given vector, the matrix elements $F_{n'n}$ can be pre-computed to get optimal performance, but complexity of the matrix multiplication remains of the order of $\mathcal{O}(N^2)$. In 1965, James W. Cooley and John Tukey [56] showed a simple way to split the computation of one DFT in two, one of the even elements of the input vector and one of its odd elements. The DFT is split according to

$$\sum_{n=0}^{N-1} F_{n'n} u_n = \sum_{n=0}^{N/2} F_{n'(2n)} u_{(2n)} + \sum_{n=1}^{N/2-1} F_{n'(2n+1)} u_{(2n+1)}.$$

The matrix components can be decomposed to

$$F_{n'(2n)} = \exp\left(-i\frac{2\pi}{N}n'(2n)\right) = \exp\left(-i\frac{2\pi}{N/2}n'n\right) = F_{n'n}^{(N/2)}$$

$$F_{n'(2n+1)} = \exp\left(-i\frac{2\pi}{N}n'(2n+1)\right)$$

$$= \exp\left(-i\frac{2\pi}{N}n'\right)\exp\left(-i\frac{2\pi}{N/2}n'n\right)$$

$$= \exp\left(-i\frac{2\pi}{N}n'\right)F_{n'n}^{(N/2)},$$

where $F^{(N/2)}$ denotes the Fourier matrix of size $N/2$. This shows, that one matrix multiplication with $F_{n'n}$ can be replaced with two matrix multiplications with $F_{n'n}^{(N/2)}$. The latter scenario is computationally more efficient and the method can be recursively applied by splitting the matrix multiplication with $F_{n'n}^{(N/2)}$ again in half. Effectively, this results in a reduced complexity of about $\mathcal{O}(N \cdot \log N)$.

In practice, the fast Fourier transforms (FFT) can reduce computation time by several orders of magnitude, compared to the standard DFT. For N that are not a power of two or even prime, suitable algorithms have been demonstrated as well (see e.g. [57]).

2.1.3 Fourier transforms on non-equispaced nodes

Evaluation of FD-OCT data involves an (inverse) Fourier transform of data which were not sampled on an equidistant scale. Instead of $\tilde{U}(k)$ only a related signal $\tilde{U}(t)$ is directly available with t being a monotone function of k. $U(z)$, the Fourier transform of $\tilde{U}(k)$, can be written as an integral transform of $\tilde{U}(t)$ by applying a suitable variable substitution

$$
\begin{aligned}
U(z) &= \int dk\, \tilde{U}(k(t)) \exp(-ikz) \\
&= \int dt\, \frac{dk(t)}{dt} \tilde{U}(k(t)) \exp(-ik(t)z).
\end{aligned}
\tag{2.1.7}
$$

This integral transform is referred to as Fourier transform on non-equispaced data. To perform it, the relation between t and k needs to be known. The inverse Fourier transform on non-equispaced data can be defined accordingly.

2.1.3.1 The non-equispaced DFT (NDFT)

The discretization of (2.1.7) is straight forward. Introducing the signals $\tilde{u}_{n'}^{(k)}$ and $\tilde{u}_{v'}$, discretizations of $\tilde{U}$ equispaced in k and equispaced in t, are given by

$$
\tilde{u}_{n'}^{(k)} = \tilde{U}(k_0 + n'\Delta k) \quad \text{and} \quad \tilde{u}_{v'} = \tilde{U}(k(t_0 + v'\Delta t)),
$$

respectively. The superscript (k) indicates that this data is linearly in k. A relation between their indices is readily obtained to

$$
n'(v') = \frac{k(t_0 - v'\Delta t) - k_0}{\Delta k}.
$$

The non-equispaced discrete Fourier transform is adapted from (2.1.2) to

$$
u_n = \sum_{v'=0}^{N-1} D_{v'n} \tilde{u}_{v'}, \quad D_{v'n} = \exp\left(-i\frac{2\pi}{N} n'(v') n\right),
\tag{2.1.8}
$$

where the derivative $dk(t)/dt$ has been neglected. u_n is thus computed by the non-equispaced discrete Fourier transform (NDFT), a multiplication with the non-equispaced Fourier matrix $D_{v'n}$. However, the FFT algorithm requires the equispacing of the input data in n'-space. A fast version of (2.1.8) can therefore not be created and its complexity remains $\mathcal{O}(N^2)$.

In analogy to a standard discrete Fourier transform being a development into orthogonal functions, sines and cosines, the NFFT can be interpreted as the development into sines and cosines, which are chirped by $k(t)$.

2.1.3.2 Interpolation and FFT (iFFT)

In order to improve computation time of an NDFT, the data $\tilde{u}_{\nu'}$ can be resampled to obtain an approximation of $\tilde{u}_{n'}^{(k)}$ prior to a standard DFT. The resampling step can, for example, be done by linear interpolation. In order to increase the accuracy of the approximation, oversampling by a factor α can be applied by computing the values $\tilde{u}_{n'}^{(k)}$ with their index n running proportional to the wavenumber k in the range 0 to $\alpha N - 1$. The interpolated signal is given by

$$
\tilde{u}_{n'}^{(k)} \approx \left(\tilde{u}_{\lceil \nu'(n'/\alpha) \rceil} - \tilde{u}_{\lfloor \nu'(n'/\alpha) \rfloor} \right) \tag{2.1.9}
$$

$$
\times \left(\nu'\left(\frac{n'}{\alpha}\right) - \left\lfloor \nu'\left(\frac{n'}{\alpha}\right) \right\rfloor \right) + \tilde{u}_{\lfloor \nu'(n'/\alpha) \rfloor}
$$

$$
= \tilde{u}_{\lceil \nu(n'/\alpha) \rceil} \underbrace{\left(\nu\left(\frac{n'}{\alpha}\right) - \left\lfloor \nu\left(\frac{n'}{\alpha}\right) \right\rfloor \right)}_{\text{weight right point}}
$$

$$
+ \tilde{u}_{\lfloor \nu(n'/\alpha) \rfloor} \underbrace{\left(1 - \left(\nu\left(\frac{n'}{\alpha}\right) - \left\lfloor \nu\left(\frac{n'}{\alpha}\right) \right\rfloor \right) \right)}_{\text{weight left point}},
$$

$$
\text{for all } n' \in [0; \alpha N - 1),
$$

where $\lceil \cdot \rceil$ and $\lfloor \cdot \rfloor$ denote the rounding up and down operations, respectively. The N input values $\tilde{u}_{\nu'}$ are used to create αN values $\tilde{u}_{n'}^{(k)}$, i.e. the length of the subsequent FFT is therefore increased by a factor α. Therefore the time complexity calculating the FFT increases to $\mathcal{O}(\alpha N \cdot \log \alpha N)$. The interpolation (2.1.9) can be rewritten by using $\lceil a \rceil - \lfloor a \rfloor = 1$, $\nu' - \lceil \nu' \rceil \leq 0$ and $\nu' - \lfloor \nu' \rfloor \geq 0$ as

$$
\tilde{u}_{n'}^{(k)} \approx \left(1 - \left| \nu'\left(\frac{n'}{\alpha}\right) - \left\lceil \nu'\left(\frac{n'}{\alpha}\right) \right\rceil \right| \right) \tilde{u}_{\lceil \nu(n'/\alpha) \rceil}
$$

$$
+ \left(1 - \left| \nu\left(\frac{n'}{\alpha}\right) + \left\lfloor \nu\left(\frac{n'}{\alpha}\right) \right\rfloor \right| \right) \tilde{u}_{\lfloor \nu(n'/\alpha) \rfloor}.
$$

By using the triangle-function $\triangle$ (A.1.3), it follows

$$\tilde{u}_{n'}^{(k)} \approx \triangle\left(v'\left(\frac{n'}{\alpha}\right) - \left\lfloor v'\left(\frac{n'}{\alpha}\right)\right\rfloor\right)\tilde{u}_{\lfloor v'(n'/\alpha)\rfloor}$$

$$+\triangle\left(v'\left(\frac{n'}{\alpha}\right) - \left\lceil v'\left(\frac{n'}{\alpha}\right)\right\rceil\right)\tilde{u}_{\lceil v'(n'/\alpha)\rceil}$$

$$= \sum_{j=\lfloor v'(n'/\alpha)\rfloor}^{\lceil v'(n'/\alpha)\rceil} \triangle\left(v'\left(\frac{n'}{\alpha}\right) - j\right)\tilde{u}_j = \sum_{j=0}^{N-1} \triangle\left(v'\left(\frac{n'}{\alpha}\right) - j\right)\tilde{u}_j,$$

where the last equality indicates, that the triangle function is zero for all data points, except $j = \lfloor v'(n'/\alpha)\rfloor$ and $j = \lceil v'(n'/\alpha)\rceil$. According to this relation, the linear interpolation performs a discrete convolution with the triangle function $\triangle$ (Figure 2.1.1a), resulting in an attenuation of the higher frequencies of the signal, which can be removed after the actual FFT by dividing the results by the Fourier transform of the triangle function. According to Section 2.1.2.2 the discrete Fourier transform of the triangle function can be evaluated by computing the analytic continuation to real n', denoted by $\triangle(k/\alpha)$. The Fourier transform is

$$\tilde{\triangle}(z) = \int dk\, \triangle\left(\frac{k}{\alpha}\right)e^{-ikz} = \int_{-\alpha}^{+\alpha} dk\left(1 - \left|\frac{k}{\alpha}\right|\right)e^{-ikz}$$

$$= \alpha\frac{\sin^2\left(\frac{z\alpha}{2}\right)}{\left(\frac{z\alpha}{2}\right)^2}.$$

This gives the required deconvolution filter

$$\tilde{\triangle}_n = \tilde{\triangle}\left(n\frac{2\pi}{N}\right) = \alpha\frac{\sin^2\left(\frac{\pi\alpha n}{N}\right)}{\left(\frac{\pi\alpha n}{N}\right)^2} = \alpha \cdot \text{sinc}^2\left(\frac{\alpha n}{N}\right),$$

which reverts the fall-off of the high frequencies.

The linear interpolation and the following FFT have approximately a time complexity of $\sim \mathcal{O}(\alpha N)$ and $\sim \mathcal{O}(\alpha N \cdot \log \alpha N)$, respectively. For optimal performance the weighting factors used for the linear interpolation and the values used for the deconvolution after the FFT can be precomputed. This algorithm will be referred to as iFFT α.

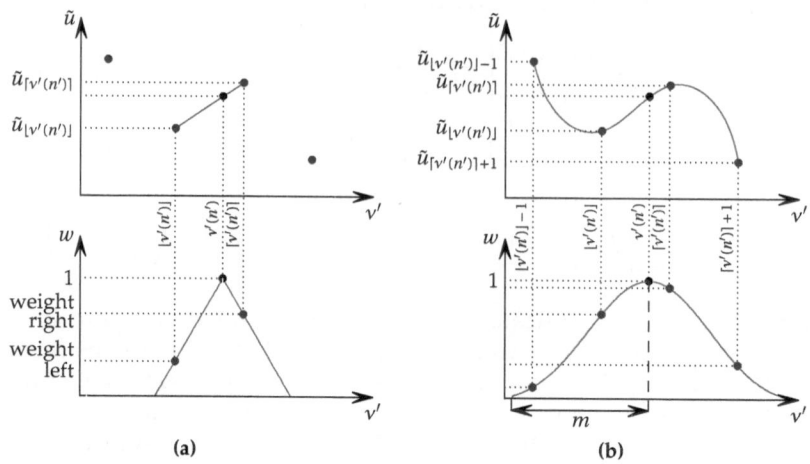

Figure 2.1.1.: Both, the interpolated FFT (a) and the non-equispaced FFT (b), perform a discrete convolution prior to the FFT to achieve interpolation. While the iFFT uses a triangle function as convolution function, the NFFT commonly uses a Kaiser-Bessel window.

2.1.3.3 The non-equispaced FFT (NFFT)

Instead of convolving with a triangle function, which involves only two original data points, an extended window function can be used which is cut off farther away from the data point that needs to be approximated (see Figure 2.1.1b) [58]. This approach is referred to as the non-equispaced fast Fourier transform (NFFT). The cut-off is determined by a parameter m and therefore the algorithm has, in addition to α, another parameter influencing accuracy and speed of the algorithm. Computation is faster if oversampling α and cut-off m are chosen smaller, but accuracy suffers and increased artifacts are observed. The discrete convolution has a complexity of about $\sim \mathcal{O}(m\alpha N)$ and therefore the complete time complexity is approximately $\sim \mathcal{O}(m\alpha N + \alpha N \cdot \log \alpha N)$.

It has been shown, that in most cases a Kaiser-Bessel window (A.1.9) yields good results [59,60] for pre-FFT interpolation. If the window function values for the data points are precomputed completely, the requirement for computing the window function does in practice not increase the time needed for the calculation of the NFFT. This algorithm will be referred to as NFFT α, m.

2.2 Sampling

2.2.1 Discretized signals

In order to apply digital signal processing to a function $U(x)$ it needs to be discretized and quantized first. No exact representation of the function $U(x)$ is known. Instead, technical devices, such as line or area cameras or photodiodes are used to acquire samples of the function $U(x)$. Lets assume the discretized or sampled function $U_s(x)$ is acquired at equidistant points $n\Delta x$ with integer n and its values at these points are identical to $U(x)$. The sampled version can be approximated by

$$
\begin{aligned}
U_s(x) &= \sum_{n=-\infty}^{+\infty} \delta(x - n\Delta x)U(x) \\
&= \mathrm{comb}\left(\frac{x}{\Delta x}\right)U(x),
\end{aligned}
$$

where the comb distribution is defined by (A.1.5).

2.2.2 The sampling theorem

The Fourier transform of the comb-distribution is again a comb-distribution, given by (A.3.8) and it thus follows that

$$
\tilde{U}_s(k) = \frac{2\pi}{\Delta x} \mathrm{comb}\left(\frac{\Delta x}{2\pi} \cdot k\right) * \tilde{U}(k). \tag{2.2.1}
$$

In the frequency domain the function $\tilde{U}(k)$ is thus repeated with a period length of $2\pi/\Delta x$, and the periods might overlap. If one assumes that the original function U is bandlimited with a cut-off frequency K

$$
\tilde{U}(k) = 0 \quad \text{for } |k| > K,
$$

this overlapping can be prevented, if it is assured that

$$
K < \frac{1}{2}\frac{2\pi}{\Delta x} = \frac{\pi}{\Delta x}, \tag{2.2.2}
$$

which is called the Nyquist-Shannon sampling criterion. The largest acquirable frequency $\pi/\Delta x$ is called Nyquist frequency. The criterion states, that a signal can only be sampled without loss of information, if the spacing frequency is at least twice the band-frequency of the signal. If it is not fulfilled, artifacts occur due to the overlapping of the frequency signals. In fact, if the criterion is fulfilled, the complete original signal $U(x)$ can be obtained by cutting off the repeated signals by multiplying with a rectangular function in the frequency domain, i.e.

$$
\begin{aligned}
\text{rect}\left(\frac{k}{2K}\right)\tilde{U}_s(k) &= \text{rect}\left(\frac{k}{2K}\right)\left(\frac{2\pi}{\Delta x}\text{comb}\left(\frac{\Delta x}{2\pi}\cdot k\right)*\tilde{U}(k)\right) \\
&= \tilde{U}(k).
\end{aligned}
$$

Going back to position space by inverse Fourier transforming, the result gives the so-called Shannon-Whittaker interpolation formula

$$
U(x) = 2K\,\text{sinc}\left(\frac{K}{\pi}x\right)*U_s(x).
$$

Interpolation with a sinc-function recreates an originally bandlimited signal. Two restrictions of the sampling theorem are important to note as they do not hold in practice: First, the sampling needs to take place over an infinite amount of sampling points, which requires an infinite amount of data. Instead, in practice the signal is acquired for a finite number of data points, making the acquisition spacelimited. A signal cannot be both, space and bandlimited and therefore the exact acquisition of a bandlimited signal is not possible. The second restriction for practical applications is that data is most often not sampled over an infinitely small amount of time or space, but rather an extended period, for example defined by the pixel size of a device or the integration time of a camera.

In the case of rectangular sampling the signal is given by

$$
\begin{aligned}
U_s(x) &= \sum_n \delta(x-n\Delta x)\int dx'\,\text{rect}\left(\frac{x'-x}{\delta x}\right)U(x') \\
&= \text{comb}\left(\frac{x}{\Delta x}\right)\cdot\left(\text{rect}\left(\frac{x}{\delta x}\right)*U(x)\right),
\end{aligned}
$$

where δx is the integration width of the sampling. In the Fourier-domain this gives

$$
\tilde{U}_s(k) = \frac{2\pi}{\Delta x}\text{comb}\left(\frac{\Delta x}{2\pi}\cdot k\right)*\left(\delta x\cdot\text{sinc}\left(\frac{\delta x}{2\pi}\cdot k\right)\cdot\tilde{U}(k)\right).
$$

If $\tilde{U}(k)$ is bandlimited, it remains so after multiplication with the sinc function, and thus the original signal can be recreated, albeit multiplied in frequency space with the sinc-function. At the Nyquist frequency $k = \pi/\Delta x$, the restored signal will be attenuated by a factor $\text{sinc}(\delta x/(2\Delta x))$. In case of a a fill-factor of 1, i.e. $\delta x = \Delta x$, the signal will be decreased by a factor 0.64 close to the Nyquist frequency.

2.2.2.1 Aliasing

Acquired signals, as well as discretized functions, can be sampled incorrectly, i.e. their signal spacing does not fulfill the Nyquist criterion (2.2.2). The comb-function in (2.2.1) causes a repetition of an actual signal in its Fourier conjugated space. If these signals overlap one speaks of aliasing, the overlapping signals cannot be separated. Aliasing can occur in spatial and in frequency domain. In OCT, this results in most cases in ghost images that closely resemble the original signal, but are oftentimes distorted, overlap the original signal and reduce signal quality.

2.3 Scalar waves

A scalar wave $U(x,t)$ is a space and time-dependent function that solves the wave equation

$$\nabla^2 U(x,t) - \frac{1}{c^2}\frac{\partial^2}{\partial t^2} U(x,t) = 0. \tag{2.3.1}$$

The wave equation is derived for the electric scalar and magnetic vector potentials and for the electric and magnetic field vector components directly from Maxwell's electrodynamic theory, if a charge and current free space is assumed. In this case c is the speed of light and the wave equation is the basis of the theory of electromagnetic radiation in general and of light and optics in particular. The precise treatment is broadly covered in the literature (see e.g. [49,51]) and will not be considered here. Additionally, only scalar wave theory will be covered, as all effects of interest in this thesis can be sufficiently and more easily described by scalar fields.

In general the field vectors of visible and infrared light cannot be measured directly. Instead its intensity I as the measurable quantity is defined by

$$I(x) = \gamma \langle (U^*U)(x,t) \rangle_t, \tag{2.3.2}$$

where $\langle \cdot \rangle_t$ denotes time-averaging and γ is a suitable conversion factor. The time-average here requires the signals to be stationary, i.e. their statistical properties do not change with time.

2.3.1 Monochromatic waves

Assuming the time dependency of the scalar wave $U(x,t)$ is only harmonic with a fixed angular frequency ω, the following approach to solve the wave equation can be used:

$$U(x,t) = U(x)e^{-i\omega t}. \tag{2.3.3}$$

Inserting (2.3.3) into (2.3.1) and introducing the free space wavenumber

$$k = \frac{\omega}{c},$$

the time-independent part of the wave $U(x)$ needs to solve the so-called Helmholtz equation

$$\nabla^2 U(x) + k^2 U(x) = 0. \tag{2.3.4}$$

The frequency of visible light determines its color and hence waves given by (2.3.3) are said to be monochromatic (from ancient Greek μονόχρωμος with μόνος meaning "one" and χρῶμα meaning "color"). In homogeneous media they have in addition to their fixed frequency also a fixed wavenumber k.

The intensity (2.3.2) of the monochromatic wave field simplifies to

$$I(x) = \gamma \langle (U^*U)(x,t) \rangle_t = \gamma \left\langle e^{-i\omega t} e^{+i\omega t} (U^*U)(x) \right\rangle_t = \gamma (U^*U)(x). \tag{2.3.5}$$

Hence, to compute the measurable intensity I of monochromatic light the time average does not need to be explicitly computed, it is known analytically. This simplification can also be used when measuring the light spectrally resolved.

2.3.2 Diffraction and propagation

Both, diffraction and propagation of a monochromatic optical field are described by the solution of the wave equation for a certain boundary condition. In the latter case the boundary condition is given by the amplitude and the phase of the propagating wave field itself, whereas in the former case this boundary condition is altered by a screen or an obstacle in the wave field. Both are described by the same mathematical formalism, and therefore the terms will be used interchangeably.

2.3.2.1 Diffraction and propagation in Fourier space

In many situations physical properties of an optical wave can be better described in frequency space, while in other situations a description in position space is preferred. Diffraction and propagation of a monochromatic optical field, i.e. finding the solution of the wave equation for a boundary condition, is easily done in frequency space compared to the long and tedious derivations of the diffraction integrals in position space.

Let the wave field U be given for a plane $z = z_0$, which means $U(x, y, z_0)$ is known and defines the boundary condition. The solution is supposed to fulfill the Helmholtz equation

$$\left(\frac{\partial^2}{\partial x^2} + \frac{\partial^2}{\partial y^2} + \frac{\partial^2}{\partial z^2} + k^2 \right) U(x, y, z) = 0. \tag{2.3.4}$$

Inserting the two-dimensional inverse Fourier transform of $\tilde{U}_{xy}$ as possible representation of U

$$U(x, y, z) = \frac{1}{(2\pi)^2} \int dx \int dy \, \tilde{U}_{xy}(k_x, k_y, z) e^{+i(k_x x + k_y y)}$$

gives a simplified version of the Helmholtz equation

$$\left(-k_x^2 - k_y^2 + \frac{\partial^2}{\partial z^2} + k^2 \right) \tilde{U}_{xy}(k_x, k_y, z) = 0.$$

The field $\tilde{U}_{xy}(k_x, k_y, z)$ is referred to as the angular spectrum of the wave field U in the plane z. With the variable k_z^2, given by $k_z^2 = k^2 - k_x^2 - k_y^2$, the problem reduces to a one-dimensional linear differential equation of second order

$$\left(\frac{\partial^2}{\partial z^2} + k_z^2 \right) \tilde{U}_{xy}(k_x, k_y, z) = 0,$$

where each pair of (k_x, k_y) defines an independent equation. The solutions to this equation are well-known from the classical harmonic oscillator:

$$\tilde{U}_{xy}(k_x, k_y, z) = A(k_x, k_y) e^{-ik_z z} + B(k_x, k_y) e^{+ik_z z}.$$

Physically, the two terms indicate forward and backward propagation. By assuming that only forward propagation is present, one can set $A(k_x, k_y) = 0$ and thus $B(k_x, k_y)$ determines the boundary condition by setting

$$B(k_x, k_y) = \tilde{U}_{xy}(k_x, k_y, z_0).$$

Therefore, diffraction can be easily described in Fourier space by a simple phase multiplication. This approach is in general referred to as the angular spectrum approach to diffraction and is frequently used in computational imaging. It has been shown, that this computation yields results equivalent to the Rayleigh-Sommerfeld integral (see e.g. [61]).

2.3.2.2 The propagator

The simple representation of propagation in Fourier space motivates the definition of the free-space propagator $\mathscr{P}$, an operator that computes the monochromatic wave field from a known plane $z = z_0$ in a distance Δz:

$$U(x, y, z_0 + \Delta z) = \mathscr{P}_{k,\Delta z}[U(x, y, z_0)], \tag{2.3.6}$$

where k is the wavenumber of the monochromatic wave. As shown above, the propagator can be described in Fourier space by

$$\tilde{U}_{xy}(k_x, k_y, z_0 + \Delta z) = P_{k,\Delta z}(k_x, k_y) \cdot \tilde{U}_{xy}(k_x, k_y, z_0).$$

Here, $P_{k,\Delta z}(k_x, k_y)$ is only a phase factor

$$P_{k,\Delta z} = e^{+ik_z \Delta z}, \tag{2.3.7}$$

with $k_z = \sqrt{k^2 - k_x^2 - k_y^2}$.

Properties   A few properties of the propagator can easily be verified mathematically and have obvious physical consequences:

- Propagation by a distance z_1 followed by a distance z_2 is identical to a propagation by a distance $z_1 + z_2$:

$$\mathscr{P}_{k,z_2} \mathscr{P}_{k,z_1} = \mathscr{P}_{k,z_1+z_2} \tag{2.3.8}$$

- Propagation by distance $\Delta z = 0$ is a neutral operation and corresponds to no change of the wave field:

$$\mathscr{P}_{k,0} = \mathrm{id} \tag{2.3.9}$$

- The complex conjugation of the propagation operator by distance z corresponds to a propagation by a distance $-z$:

$$\mathscr{P}^*_{k,z}[\cdot] = \mathscr{P}_{k,-z}[\cdot]$$

- The propagation commutes with the multiplication with a constant complex factor c:

$$c\mathscr{P}_{k,z}[\cdot] = \mathscr{P}_{k,z}[c\cdot]$$

2.3.3 Broadband light

So far only light, that is a solution to the Helmholtz equation (2.3.4) and is thus monochromatic, has been considered. The following Section will treat more general solutions, but in order to keep it simple, the attention will be restricted to one-dimensional waves and functional dependencies on x and y will be dropped.

The general solution to the one-dimensional wave equation

$$\frac{\partial^2}{\partial z^2} U(z,t) - \frac{1}{c^2}\frac{\partial^2}{\partial t^2} U(z,t) = 0$$

is

$$U(z,t) = \bar{f}(z+ct) + f(z-ct),$$

with f and $\bar{f}$ being arbitrary functions, which is shown by applying the chain rule of differentiation. Restricting the attention to forward propagation by setting $\bar{f} = 0$, one can use the Fourier transform of the solution

$$\tilde{f}(k) = 2\pi \int d(z-ct) f(z-ct) e^{-i(z-ct)k}$$

to express $U(z,t)$ by a superposition of harmonic waves with the amplitude $\tilde{f}(k)$

$$U(z,t) = \int dk\, \tilde{f}(k) e^{+ikz - \omega(k)t}, \tag{2.3.10}$$

where the angular frequency $\omega(k) = ck$ has been used. The absolute value of $\tilde{f}$ contains the spectral shape of the wave and determines its coherence properties. The phase of $\tilde{f}$ contains information on the temporal shape of the wave.

In order to take spectral dependence and overall amplitude into account, one can separate the function $\tilde{f}(k)$ in a normalized complex amplitude spectrum $A_C(k)$ and an overall amplitude A, where $A_C(k)$ is normalized by the intensity spectrum

$$S(k) = |A_C|^2(k), \quad \int dk\, S(k) = 1.$$

By substituting variables and exchanging k by ω/c, the wave (2.3.10) can be written as

$$U(z,t) = \frac{A}{c} \int d\omega\, e^{-i\omega t} A_C\left(\frac{\omega}{c}\right) e^{+i\frac{\omega}{c}z}. \tag{2.3.11}$$

The intensity I as defined by (2.3.2) becomes a simple expression

$$I(z) = \gamma \langle (U^*U)(z,t) \rangle_t$$

$$\propto \lim_{T \to \infty} \int_{-T/2}^{T/2} dt \, \frac{A}{c} \int d\omega' \, e^{i\omega' t} A_C^* \left(\frac{\omega'}{c} \right) e^{-i\frac{\omega'}{c} z}$$

$$\times \frac{A}{c} \int d\omega \, e^{-i\omega t} A_C \left(\frac{\omega}{c} \right) e^{+i\frac{\omega}{c} z}$$

$$= \frac{2\pi A^2}{c^2} \int d\omega \, |A_C|^2 \left(\frac{\omega}{c} \right) = \frac{2\pi A^2}{c} \int dk \, |A_C|^2(k). \qquad (2.3.12)$$

The division by the overall time T to build the time-average has been dropped. A stationary or ergodic wave field, as required for the a proper statistical definition of this time-average, is in general not Fourier transformable and the Fourier integral does not converge. Here only the proportionality is considered, which gives sufficient accurate results; a more mathematical rigorous approach to this problem can e.g. be found in [51, 62].

For visible and near-infrared light, the frequency $\omega = 2\pi c/\lambda$ is in the range between 10^{14} Hz to 10^{16} Hz. The usual data acquisition is not faster than a few GHz, justifying the approximation to only consider the infinitely, stationary time-averaged signal.

2.3.3.1 Dispersion

In materials the speed of light c is reduced by the refractive index n of the material. Since the response of the material to the oscillating electric and magnetic fields is frequency dependent also the refractive index depends on the frequency of the passing wave. To describe this effect, known as (chromatic) dispersion, the constant refractive index n is replaced by a frequency dependent function $n(\omega)$, yielding the modified wave equation

$$\frac{\partial^2}{\partial z^2} U(z,t) - \frac{n^2(\omega)}{c^2} \frac{\partial^2}{\partial t^2} U(z,t) = 0. \qquad (2.3.13)$$

The solution to this equation is in analogy to (2.3.11) given by

$$U(z,t) = \frac{A}{c} \int d\omega \, e^{-i\omega t} A_C \left(\frac{\omega}{c} \right) e^{+ik(\omega)z}, \qquad (2.3.14)$$

where the dispersion relation

$$k(\omega) = \frac{n(\omega)}{c}\omega$$

needs to hold.

For a monochromatic wave with $\omega = \omega_0$, one can introduce the solution $U(z,t) = U(z)e^{-i\omega_0 t}$, inserting it in (2.3.13) gives a modified Helmholtz equation

$$\frac{\partial^2}{\partial z^2}U(z) + \frac{n^2(\omega_0)\omega_0^2}{c^2}U(z) = 0.$$

Its solution is

$$U(z) = AA_C e^{+ik(\omega_0)z},$$

where $k(\omega_0) = n(\omega_0)\omega_0/c$. By introducing the optical path length

$$z' = z \cdot n$$

and the free-space wavenumber $k_0 = \omega_0/c$, it can be rewritten to

$$U(z'/n) = AA_C e^{+ik_0 z'(k_0)},$$

where optical path length z' depends on the wavenumber k_0.

2.3.3.2 Group and phase velocity

More insight into the behavior of broadband light in the case of dispersion is gained by developing the dispersion relation in a Taylor series around the central frequency ω_0

$$k(\omega) = k(\omega_0) + \frac{dk(\omega)}{d\omega}\bigg|_{\omega=\omega_0}(\omega - \omega_0) + \mathcal{O}\big((\omega-\omega_0)^2\big). \tag{2.3.15}$$

Inserting this into (2.3.14) gives

$$
\begin{aligned}
U(z,t) &= \frac{A}{c}\int d\omega\, e^{-i\omega t} A_C\left(\frac{\omega}{c}\right) e^{+ik(\omega_0)z+i\frac{dk(\omega)}{d\omega}\big|_{\omega=\omega_0}(\omega-\omega_0)z} \\
&= \underbrace{\frac{A}{c}e^{-i\omega_0 t+ik(\omega_0)z}}_{\text{phase-term}} \cdot \underbrace{\int d\omega\, A_C\left(\frac{\omega}{c}\right) e^{-i(\omega-\omega_0)\left(t+\frac{dk(\omega)}{d\omega}\big|_{\omega=\omega_0}z\right)}}_{\text{envelope-term}}.
\end{aligned}
$$

In this expression it becomes obvious that the first term will only influence the overall phase of the wave, as its amplitude is constant. The second term completely determines the instantaneous amplitude and therefore the energy of U; the t and z dependence of this second term entirely describe the motion of the envelope. A time and space transform by

$$t \quad \rightarrow \quad t' = t + \delta t = t - \left.\frac{dk(\omega)}{d\omega}\right|_{\omega=\omega_0} \delta z$$

$$z \quad \rightarrow \quad z' = z + \delta z,$$

leaves the envelope term invariant. It follows the group velocity, the velocity of points of constant amplitude

$$\frac{1}{v_g} = -\frac{\delta t}{\delta z} = \left.\frac{dk(\omega)}{d\omega}\right|_{\omega=\omega_0}. \tag{2.3.16}$$

Quadratic and higher order terms of the dispersion relation (2.3.15) are referred to as group velocity dispersion (GVD), as they cause the group velocity (2.3.16) to depend on the frequency ω_0. The phase-term determines the velocity of phase fronts in the wave, called phase-velocity v_p. It is given by

$$v_p = \frac{\omega_0}{k(\omega_0)} = \frac{c}{n(\omega_0)}.$$

2.3.4 Coherence

Coherence is a property of a wave field, that describes the correlation of the field at different points in time and space. At two space-time points (x_1, t_1) and (x_2, t_2) the degree of coherence of a wave field U with respect to these points can be described by the coherence function

$$\Gamma(x_1, t_1; x_2, t_2) = \langle U^*(x_1, t_1) U(x_2, t_2) \rangle_t,$$

where the field is assumed stationary, i.e. the statistical properties of the wave field do not change with time. If only the statistical properties are of interest, the function can be normalized

$$\gamma(x_1, t_1; x_2, t_2) = \frac{\Gamma(x_1, t_1; x_2, t_2)}{\sqrt{\Gamma(x_1, t_1; x_1, t_1)\Gamma(x_2, t_2; x_2, t_2)}}$$

$$= \frac{\Gamma(x_1, t_1; x_2, t_2)}{\sqrt{\langle (U^*U)(x_1, t_1) \rangle_t \langle (U^*U)(x_2, t_2) \rangle_t}},$$

where the function γ is called the degree of coherence. Two special cases of the general coherence property are often distinguished. The degree of coherence at two different points of the wave field at the same time is called spatial coherence; the degree of coherence of a single point in space at different times is called the temporal coherence. The temporal coherence depends strongly on the intensity spectrum of the light source, and will be investigated further in the next Section.

2.3.4.1 Temporal coherence

Restricting the attention to temporal coherence, one can evaluate the temporal degree of coherence by

$$\gamma\left(\Delta t = \frac{z}{c}\right) = \frac{\langle U^*(z_0, t) U(z_0, t + \Delta t) \rangle_t}{\left\langle |U|^2(z_0, t) \right\rangle_t}.$$

Using a calculation analogous to the derivation of (2.3.12), this yields

$$\gamma\left(\Delta t = \frac{z}{c}\right) = \frac{\int d\omega \, |A_C|^2 \left(\frac{\omega}{c}\right) e^{+i\omega\Delta t}}{\int d\omega \, |A_C|^2 \left(\frac{\omega}{c}\right)} = \int d\omega \, |A_C|^2 \left(\frac{\omega}{c}\right) e^{+i\omega\Delta t}. \qquad (2.3.17)$$

The coherence time τ and the coherence length $L = c\tau$ are defined by the width of the coherence function, i.e.

$$\gamma(\tau) = \gamma\left(\frac{L}{c}\right) = \frac{1}{2}.$$

The temporal degree of coherence describes the normalized interference signal of a Michelson interferometer. Given is therefore a Michelson interferometer with a broadband light source as shown in Figure 2.3.1. The light in one arm passes a length z_0 and in the other arm a length $z_0 + 2z$, before reaching the detector, if the mirror in the second arm is displaced by z. The fields at the detector of the two arms can be described by

$$U_1(t) = \frac{A_1}{c} \int d\omega \, e^{-i\omega t} A_C\left(\frac{\omega}{c}\right) e^{i\frac{\omega}{c} z_0}$$

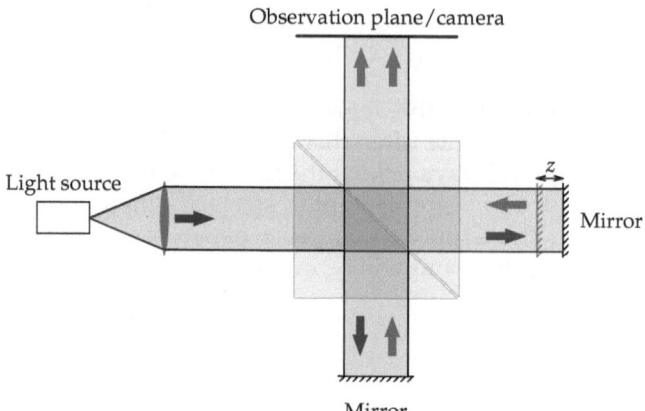

Figure 2.3.1.: Schematic drawing of a Michelson interferometer. By default, the light needs to travel the same optical path length z_0 in both interferometer arms to reach the camera/observer. If one mirror is displaced by an additional distance z, the additional optical path length for this arm becomes $2z$.

and

$$U_2(t) = \frac{A_2}{c} \int d\omega' \, e^{-i\omega' t} A_C\left(\frac{\omega'}{c}\right) e^{i\frac{\omega'}{c}(z_0 + 2z)}.$$

Superimposing the light of both arms gives the intensity

$$
\begin{aligned}
I &= \gamma \langle |U_1 + U_2|^2 \rangle_t \\
&= \underbrace{\gamma \langle U_1^* U_1 \rangle}_{I_1} + \underbrace{\gamma \langle U_2^* U_2 \rangle_t}_{I_2} + \gamma 2 \operatorname{Re}\langle U_1^* U_2 \rangle_t.
\end{aligned}
$$

According to (2.3.12) the first two terms will evaluate to the light intensities of the interferometer arms. The third term evaluates to

$$
\begin{aligned}
\langle U_1^*(t) U_2(t) \rangle_t &\propto \lim_{T \to \infty} \int_{-T/2}^{T/2} dt \, \frac{A_1}{c} \int d\omega \, e^{i\omega t} A_C^*\left(\frac{\omega}{c}\right) e^{-i\frac{\omega}{c} z_0} \\
&\quad \times \frac{A_2}{c} \int d\omega' \, e^{-i\omega' t} A_C\left(\frac{\omega'}{c}\right) e^{i\frac{\omega'}{c}(z_0 + 2z)} \\
&= \frac{2\pi A_1 A_2}{c^2} \int d\omega \, |A_C|^2\left(\frac{\omega}{c}\right) e^{i\frac{\omega}{c} 2z}.
\end{aligned}
\tag{2.3.18}
$$

As explained in Section 2.3.3, only proportionality is considered here and the factor $1/T$ in the time average has been omitted. The measured interference amplitude as function of the displacement of the two mirrors is proportional to the Fourier transform of the intensity spectrum $|A_C|^2(k)$. This is the basis of time-domain optical coherence tomography, which will be explained in more detail in Section 2.7.1.

2.4 Coherent imaging

2.4.1 Image formation

Given are the intensity distributions of the light field $I_{obj}(x,y)$ and $I_{image}(x,y)$ in the two planes $z = z_{obj}$ and $z = z_{image}$, called the object and image plane, respectively. Imaging is the process of mapping the light intensity distribution in the object plane to the distribution in the image plane, ideally given by

$$I_{image}(x,y) = \frac{1}{M^2} I_{obj}\left(\frac{x}{M}, \frac{y}{M}\right), \tag{2.4.1}$$

where M is called the magnification of the imaging system and the factor $1/M^2$ is required for energy conservation. The aim of imaging is to bring all light originating from a specific lateral position in the object plane to a specific lateral position in the image plane, and to maintain relative distances between these lateral positions. The mapping is done by using lenses and other optical devices, but usually this cannot achieve an ideal image (2.4.1).

Usually, imaging systems are linear systems, i.e. they obey the superposition principle. In coherent imaging, for spatially and temporally coherent waves, the waves superimpose and interference occurs: The system is linear in its complex wave fields. In incoherent imaging, the interferences cancel and the intensities superimpose. Here, only the coherent case will be considered and the respective imaging systems are described by integral transforms $\mathscr{I}[\cdot]$ of wave fields U from object to image plane. The intensity in the image plane is then given by

$$
\begin{aligned}
I_{image}(\bar{x},\bar{y}) &= U_{image}^*(\bar{x},\bar{y})U_{image}(\bar{x},\bar{y}) \\
&= \mathscr{I}\big[U_{obj}(x,y)\big]^* \mathscr{I}\big[U_{obj}(x,y)\big],
\end{aligned}
$$

where U_{image} is the wave field in the image plane and U_{obj} is the wave field in the object plane and thus $I_{obj}(x,y) = \left(U_{obj}^* U_{obj}\right)(x,y)$. For convenience, the lateral coordinates in the image plane have been scaled according to the magnification of the imaging system, i.e.

$$\bar{x} = Mx, \quad \bar{y} = My.$$

The linearity of the imaging system, following from the superposition principle of waves, can be written as

$$
\begin{aligned}
aU_{\text{image},1}(\bar{x},\bar{y}) + bU_{\text{image},2}(\bar{x},\bar{y}) &= a\mathscr{I}\Big[U_{\text{obj},1}(x,y)\Big] + b\mathscr{I}\Big[U_{\text{obj},2}(x,y)\Big] \\
&= \mathscr{I}\Big[aU_{\text{obj},1}(x,y) + bU_{\text{obj},2}(x,y)\Big],
\end{aligned}
$$

with arbitrary image and object fields $U_{\text{image},1}$, $U_{\text{image},2}$, $U_{\text{obj},1}$, and $U_{\text{obj},2}$. Writing the object field as superposition of point sources

$$U_{\text{obj}}(x,y) = \int dx'\, dy'\, U_{\text{obj}}(x',y')\delta^{(2)}(x'-x,y'-y),$$

and using the superposition principle to determine the corresponding image field, it follows that

$$
\begin{aligned}
U_{\text{image}}(\bar{x},\bar{y}) &= \mathscr{I}\Big[\int dx'\, dy'\, U_{\text{obj}}(x',y')\delta^{(2)}(x'-x,y'-y)\Big] \\
&= \int dx'\, dy'\, U_{\text{obj}}(x',y')\mathscr{I}\Big[\delta^{(2)}(x'-x,y'-y)\Big],
\end{aligned}
$$

i.e. the imaging system is completely determined by its response to all possible point sources at all lateral positions. An additional assumption that holds, at least approximately for real imaging, is translational invariance

$$U_{\text{image}}(\bar{x}+\Delta\bar{x},\bar{y}+\Delta\bar{y}) = \mathscr{I}\Big[U_{\text{obj}}(x+\Delta x,y+\Delta y)\Big], \qquad (2.4.2)$$

with $\Delta\bar{x} = M\Delta x$ and $\Delta\bar{y} = M\Delta y$ for all Δx and Δy. In this case the imaging system can be described by a single point source, the response for all other lateral positions follows from the translation law (2.4.2). A further simplification arises by introducing

$$h(x'-\bar{x},y'-\bar{y}) = \mathscr{I}\Big[\delta^{(2)}(x'-x,y'-y)\Big].$$

The integral transform can now be written as a convolution

$$U_{\text{image}}(\bar{x},\bar{y}) = \int dx'\, dy'\, U_{\text{obj}}(x',y')h(x'-\bar{x},y'-\bar{y}),$$

or, equivalently, in the Fourier-domain, defining the amplitude transfer function $\tilde{h}$, by

$$\tilde{U}_{\text{image}}(k_x, k_y) = \tilde{U}_{\text{obj}}(k_x, k_y) \cdot \tilde{h}(k_x, k_y).$$

2.4.2 Quadratic phase factors

Especially in paraxial approximation, quadratic phase factors frequently occur. For convenience, a quadratic phase factor function can be introduced:

$$Q_a(x, y) \equiv \exp\left(i\frac{a}{2}\left(x^2 + y^2\right)\right).$$

Q_a is similar to a Gaussian function, albeit with imaginary variance, its Fourier transform is again a quadratic phase factor

$$\mathscr{F}_{xy}[Q_a(x, y)] = \frac{2\pi i}{a} Q_{-1/a}(k_x, k_y).$$

For example, by using $\sqrt{1 - x^2} \approx 1 - x^2/2$, the argument of the propagator kernel $P_{k,z}$ can be approximated by a polynomial and a lateral frequency independent phase factor

$$P_{k,z}(k_x, k_y) \approx \exp(izk) \cdot Q_{-z/k}(k_x, k_y). \tag{2.4.3}$$

The two-dimensional Fourier transform of this can be computed to

$$\tilde{P}_{k,z}(x, y) \approx -\frac{ik}{2\pi z} \exp(izk) \cdot Q_{k/z}(x, y). \tag{2.4.4}$$

This approximation is valid if $x^2 \ll 1$ and thus $k_x^2/k^2 \ll 1$, i.e. if the lateral frequencies are small compared to the wavenumber. In digital computations where wave fields are acquired with a few microns pixel spacing, this approximation is valid and the error is small.

2.4.3 Thin lenses

A lens is an optical instrument that focuses all light of a plane wave onto a single spot when the light passes the lens. The single spot, the focus, is in a distance f of the lens, referred to as focal length. The lateral position of the focus thereby depends on the incident angle of the plane wave. A wave that is focused onto a specific focal point is a spherical wave, its formula can be written as

$$
\begin{aligned}
S_k(x,y,z) &= A \frac{\exp\left(-ik\sqrt{x^2+y^2+z^2}\right)}{\sqrt{x^2+y^2+z^2}} \\
&\approx A \frac{\exp\left(-i\frac{k}{2z}\left(x^2+y^2\right)+ikz\right)}{z} = \frac{A}{z}Q_{k/z}(x,y)\exp(ikz), \quad (2.4.5)
\end{aligned}
$$

if its focus is in the origin $(x,y,z)=0$. A plane wave is described by

$$
U(x,y,z) = A\exp[ikz]. \tag{2.4.6}
$$

A converging lens in the plane $z=-f$ is supposed to create the spherical wave (2.4.5), if its incident wave is given by (2.4.6), with its effect only taking place in the plane $z=-f$. Assuming an only multiplicative effect by phase factors $L_{k,f}(x,y)$ it follows

$$
S_{k,f}(x,y,-f) = L_{k,f}(x,y) \cdot U(x,y,-f),
$$

and therefore

$$
L_{k,f}(x,y) = \exp\left[-i\frac{k}{2f}\left(x^2+y^2\right)\right] = Q_{-k/f}(x,y), \tag{2.4.7}
$$

where the amplitude of the spherical wave has been neglected as lenses do not influence it. The effects of a thin lens with focal length f and of a propagation by a distance f are remarkably similar. Both are represented by phase multiplications with practically identical phase factors, albeit the former is performed in position space, while the later is performed in frequency space. A more rigorous derivation of the effects of a lens on a wave field can be found in [49,50].

2.4.4 The Fresnel approximation

By applying the convolution theorem (A.3.1) and using the Fourier-domain representation of the propagation kernel $\tilde{P}_{k,z}$, the propagation (2.3.6) can be written as a convolution

$$
U(x,y,z_0+z) = \tilde{P}_{k,z}(x,y) * U(x,y,z_0).
$$

Using the paraxial approximation of $\tilde{P}_{k,z}$, given by (2.4.4), it follows

$$Um(x,y,z_0 + z) = -\frac{ik}{2\pi z}e^{izk}$$
$$\times \int dx' \int dy' \exp\left[i\frac{k}{2z}\left((x-x')^2 + (y-y')^2\right)\right]U(x',y',z_0).$$

This approximation is the Fresnel approximation. By expanding the quadratics in the exponential and introducing scaled x and y variables

$$x \to \bar{x} = xk/z, \quad y \to \bar{y} = yk/z \tag{2.4.8}$$

the equation simplifies to

$$U(\bar{x}, \bar{y}, z_0 + z) = -\frac{ik}{2\pi z}e^{izk}Q_{z/k}(\bar{x}, \bar{y})\mathscr{F}_{xy}[Q_{k/z}(x,y)U(x,y,z_0)], \tag{2.4.9}$$

where $\mathscr{F}_{xy}[\cdot]$ is supposed to transforms from (x,y) to the conjugated $(\bar{x}, \bar{y})$-space. This computation is only a multiplication of the field with a phase factor, followed by a Fourier transform and an (additional) final phase factor multiplication. This procedure is referred to as Fresnel transform, and it is a computationally very efficient way to compute a diffracted field. Its disadvantage is, that the variable substitutions (2.4.8) depend on the propagation distance z and the wavenumber k. The Fresnel approximation is only valid on propagation distances that are large compared to the wavelength.

Inserting a wave field U, subjected to a lens L with focal length f, described by $L_{k,f}(x',y') \cdot U(x',y',z_0)$ in (2.4.9), and computing the optical wave field in the focal plane of the lens by setting $z = f$, one sees that the lens phase factor cancels with the phase factor of the Fresnel transform. It remains the Fourier transform and a phase factor applied after the transform

$$U(\bar{x}, \bar{y}, z_0 + f) = -\frac{ik}{2\pi f}\exp(ifk)Q_{z/k}(\bar{x}, \bar{y})\mathscr{F}_{xy}[U(x,y,z_0)]. \tag{2.4.10}$$

The lateral coordinates in the focal plane are here given by

$$x \to \bar{x} = xk/f, \quad y \to \bar{y} = yk/f. \tag{2.4.11}$$

When using coherent light, imaging through a lens thus gives the intensity distribution belonging to the Fourier transform of the incident wave field, however, phase factors themselves differ.

2.4.4.1 Conjugated planes

In the last Section it was shown, that the paraxial propagation, defined as a Fourier transform, followed by a quadratic phase multiplication and an inverse Fourier transform, can be represented as a single Fresnel transform after suitable coordinate transformation, comprising a single Fourier transform. A similar effect can be achieved, when considering a Fresnel transform, propagating by a distance z, followed by a lens with focal length $f = z$, which is a quadratic phase factor in position space, followed by a second Fresnel transform, again propagating by a distance z. The quadratic phase factor of the lens and the two inner phase factors of the two Fresnel transforms, together with the two Fourier transforms of the Fresnel transforms, effectively represent a convolution with a quadratic phase factor. This convolution can again be expanded in a single Fresnel transform. In this specific scenario, all phase factors cancel, leaving only the Fourier transform.

To show this, we introduce the propagated wave field before it enters the lens

$$U(x, y, z_0) = \tilde{P}_{k,f}(x, y) * U(x, y, z_0 - f),$$

and its Fourier transform

$$
\begin{aligned}
\mathscr{F}_{xy}[U(x, y, z_0)] &= P_{k,f}(\bar{x}, \bar{y}) \cdot \tilde{U}_{xy}(\bar{x}, \bar{y}, z_0 - f) \\
&= \exp(ikf)Q_{-z/k}(\bar{x}, \bar{y})\mathscr{F}_{xy}[U(x, y, z_0 - f)].
\end{aligned}
$$

Inserting in (2.4.10) gives

$$U(\bar{x}, \bar{y}, z_0 + f) = -\frac{ik}{2\pi f} \exp(ik2f)\mathscr{F}_{xy}[U(x, y, z_0 - f)]. \tag{2.4.12}$$

Except for a constant x and y-independent factor the field in one focal plane of a collective lens is the Fourier transform of the field in the other focal plane and vice versa.

2.4.4.2 Relation to the angular spectrum

Equation (2.4.12) states, that the effect of a lens is closely related to a Fourier transform and consequently a close relation to the angular spectrum exists. Considering a plane wave U with propagation direction $\hat{k} = \left(\hat{k}_x, \hat{k}_y, \hat{k}_z\right)$, $\left\|\hat{k}\right\| = 1$ and wavenumber k incident on the lens. The plane wave in the focus of a lens at $z = z_0$, i.e. in the plane $z = z_0 - f$, is given by

$$U(x, y, z_0 - f) = AA_C e^{ik\left(\hat{k}_x x + \hat{k}_y y + \hat{k}_z (z_0 - f)\right)}.$$

Inserting in (2.4.12) gives the field in the focal plane

$$U(\bar{x}, \bar{y}, z_0 + f) = -\frac{ik}{2\pi f} A A_C \exp\left(ik\left(f + \hat{k}_z z_0\right)\right) \delta\left(\bar{x} - k\hat{k}_x, \bar{y} - k\hat{k}_y\right).$$

Of interest is the resulting δ-distribution, it ensures with (2.4.11) that

$$\bar{x} = \frac{xk}{f} = k\hat{k}_x \quad \text{and} \quad \bar{y} = \frac{yk}{f} = k\hat{k}_y.$$

It follows the relation between incident direction of the plane wave and lateral focusing position:

$$x = f\hat{k}_x = f\frac{k_x}{k} \quad \text{and} \quad y = f\hat{k}_y = f\frac{k_y}{k}.$$

A lens therefore translates directions to lateral positions and vice versa.

2.4.5 Lateral resolution

2.4.5.1 The point spread function

The quality of an imaging system can well be characterized using the response to a single delta peak. We therefore introduce the intensity point spread function (PSF) as the intensity distribution in the image plane with a single point source in the object plane:

$$p(\bar{x}, \bar{y}) = \mathscr{I}\left[\delta^{(2)}(x,y)\right]^* \mathscr{I}\left[\delta^{(2)}(x,y)\right].$$

2.4.5.2 The Rayleigh criterion

Resolution is commonly interpreted as the minimum distance two point sources of equal intensity in the object plane need to have, in order to be distinguishable in the image plane. For coherent imaging a problem arises, as this distance depends on the relative phase of the two waves: if their overlapping areas interfere destructively they are better distinguished than in the case of constructive interference.

Consequently, lots of different criteria exist to define resolution (see e.g. [51]). In incoherent imaging systems, with sharp-edged apertures, the Rayleigh criterion is used widely. The Rayleigh criterion defines the resolution as the distance of the PSFs maximum to its nearest minimum. It should be noted that the Rayleigh criterion is not suitable in many situations, e.g. when using apodized beams, Gaussian beams or coherent imaging. Still an equivalent condition can be found for coherent imaging, by computing the distance of two point sources required, to give the same contrast as obtained using the Rayleigh criterion for incoherent imaging.

2.4.5.3 The numerical aperture (NA)

According to the uncertainty relation, a small point in the position space wave field, requires a large structure in angular spectrum of the wave field. Resolution is therefore determined by the width of the angular spectrum amplitudes. This criterion is independent of the wavelength, it so far only involves the lateral components of the angular spectrum k_x and k_y, not its absolute value.

The angular spectrum of a wave field describes the propagation directions of the wave field. If certain propagation directions cannot contribute to the process of image formation and parts of the angular spectrum are cut, the resolving power is limited and imaging artifacts occur. In general, certain propagation directions do not contribute because of geometrical reasons: The size of lenses and other pupils and apertures are limited and thus not all rays can pass them.

The lateral components of the k-vector of the angular spectrum are only proportional to the propagation direction as long as the absolute wavenumber is fixed. The resolving power of an imaging system therefore depends on the wavenumber.

Considering a single lens with focal length f and radius r, we can geometrically approximate the angular spectrum that passes this lens:

$$\left| \frac{k_x}{k} \right| \lesssim \frac{r}{\sqrt{f^2 + r^2}}.$$

As shown in Figure 2.4.1 this is the sine of the half opening angle $\theta/2$ of the objective and motivates the definition of the Numerical Aperture (NA)

$$\text{NA} = n \cdot \sin \frac{\theta}{2},$$

with n being the refractive index of the medium that is considered. Using the uncertainty relation $\sigma_{k_x} \sigma_x \gtrsim 1/2$, we can also estimate the resulting resolution

$$\sigma_x \gtrsim \frac{n}{2k \cdot \text{NA}} = \frac{n\lambda}{4\pi \cdot \text{NA}},$$

where σ_{k_x} denotes the standard-deviation of the x-component of the angular spectrum and σ_x the standard-deviation of the x-position coordinate, a measure for the lateral resolution. These definitions are in general not in accordance with the

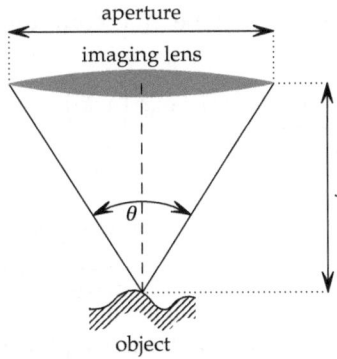

Figure 2.4.1.: The numerical aperture (NA) depends on the angle in which light of the sample can enter the imaging system. It is determined by the pupil diameter and the focal length of the objective lens. The higher the NA, the higher the lateral resolution.

Rayleigh-criterion or the exact definition of the aperture, but they give strong upper limits and show correct proportionality. The exact formula depends on the shape of the aperture, and a few solutions are found in Appendix B.2.

2.4.5.4 Gaussian beams

A Gaussian function has the special property that its Fourier transform is a Gaussian function itself. Upon propagation, the intensity will always keep a Gaussian shape. In optics the Gaussian beam in its focus is defined by its amplitude function

$$U(x, y, z_0) = A_0 \exp\left(-\frac{x^2 + y^2}{w_0^2}\right), \qquad (2.4.13)$$

where w_0 is the $1/e^2$-intensity width of the beam in its focal plane $z = z_0$. As shown in Appendix B.2.3, by using the angular spectrum approach in paraxial approximation, the field of a Gaussian beam is computed to

$$U(r, z_0 + z) = A_0 \frac{w_0}{w(z)} \underbrace{\exp(-i\zeta(z))}_{\text{phase-shift at } z = 0} \underbrace{\exp\left(izk\right)}_{\text{overall phase}}$$

$$\times \underbrace{\exp\left(ik\frac{r^2}{2R(z)}\right)}_{\text{wave-front curvature}} \underbrace{\exp\left(-\frac{r^2}{w^2(z)}\right)}_{\text{local beam width}},$$

with $r = \sqrt{x^2 + y^2}$. The first exponential term is a phase-shift in the focus, known as the Gouy phase shift. Its argument is defined as

$$\zeta(z) = \arctan\left(\frac{z}{z_R}\right),$$

with the Rayleigh length z_R defined as

$$z_R = \frac{1}{2}kw_0^2 = \frac{\pi w_0^2}{\lambda}. \qquad (2.4.14)$$

The second exponential term gives the overall phase change and is comparable to an infinite plane wave. In fact for $w_0 \to \infty$, the Rayleigh length z_R approaches infinity and the other exponential terms vanish. The third exponential term gives an additional phase, dependent on the lateral position r and thus determines the curvature of the wave-front. The radius of curvature function is given by

$$R(z) = \frac{z^2 + z_R^2}{z} = z\left(1 + \frac{z_R^2}{z^2}\right).$$

The fourth and last exponential term is a Gaussian and determines the local beam width.

$$w^2(z) = \frac{2}{k}\frac{z^2 + z_R^2}{z_R} = \frac{2z_R}{k}\left(\frac{z^2}{z_R^2} + 1\right) = w_0^2\left(1 + \frac{z^2}{z_R^2}\right).$$

In the according intensity distribution $I(r, z)$ the complex parts of this equation cancel.

$$I(r, z) = \gamma(U^*U)(r, z) = \gamma A_0^2 \frac{w_0^2}{w^2(z)}\exp\left(-\frac{2r^2}{w^2(z)}\right).$$

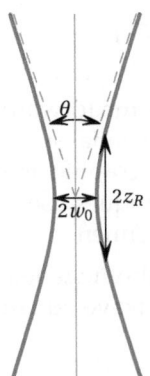

Figure 2.4.2.: Schematic drawings of a Gaussian beam. The $1/e^2$ intensity boundaries are shown (red) together with the divergence angle θ, the beam width w_0 in the focus, and the Rayleigh length z_R.

A schematic drawing of such a Gaussian beam is shown in Figure 2.4.2.

The NA of a Gaussian beam is not well defined as the aperture of the beam does not have sharp boundaries. Most commonly the $1/e^2$ intensity-width far away from the focus is being used to define the focusing angle θ:

$$\frac{\theta}{2} \approx \frac{w(z)}{\sqrt{z^2 + w^2(z)}}, \quad \text{for } z \gg z_R.$$

Asymptotically, i.e. for large z it follows

$$\frac{z_R^2}{w_0^2} = \frac{n^2}{NA^2} - 1.$$

Accordingly, formulas for the central beam width w_0 and the Rayleigh length z_R can be obtained:

$$w_0 = \frac{2}{k}\sqrt{\frac{n^2}{NA^2} - 1} \approx \frac{\lambda_0}{\pi NA},$$

$$z_R = \frac{2}{k}\left(\frac{n^2}{NA^2} - 1\right) \approx \frac{\lambda_0}{\pi NA^2} = \frac{w_0}{NA},$$

with λ_0 being the vacuum wavelength; the approximations are only valid for small NA.

2.5 Optical scattering theory

Optical scattering theory describes the forward and backscattering of (monochromatic) light from a medium with a spatial structure defined by the refractive index $n(x)$. It is the basis of optical tomographic imaging techniques that are based on scattering, however, different techniques use different methodologies to measure the scattering properties of the specimen.

The cause of scattering is the inhomogeneity of the refractive index. In order to describe it mathematically, the wave equation is modified to have a position dependent refractive index n:

$$\nabla^2 U(x,t) - \frac{n^2(x)}{c^2} \frac{\partial^2}{\partial t^2} U(x,t) = 0. \tag{2.5.1}$$

This wave equation does no longer follow from Maxwell's equations for the electric and magnetic field vector components. Nevertheless, it is for most practical situations a very good approximation (see e.g. [51]). In analogy to Section 2.3.1, we introduce the time-independent wave function $U(x)$

$$\nabla^2 U(x) + k^2 n^2(x) U(x) = 0.$$

2.5.1 The Green's function

For a mathematical analysis of the inhomogeneous wave equation (2.5.1), the Green's function $G(x)$ of the Helmholtz equation is required. It is defined by

$$\nabla^2 G(x) + k_0^2 G(x) = \delta^{(3)}(x), \tag{2.5.2}$$

where $\delta^{(3)}$ is the three-dimensional δ-distribution. This function can be most easily derived in Fourier space, by using the approach

$$G(x) = \frac{1}{(2\pi)^3} \int d^3k \, \tilde{G}(k) e^{+ik \cdot x} \tag{2.5.3}$$

and using the Fourier representation of the δ-distribution

$$\delta^{(3)}(x) = \frac{1}{(2\pi)^3} \int d^3k \, e^{+ik \cdot x}. \tag{2.5.4}$$

Inserting (2.5.3) and (2.5.4) in (2.5.2) gives

$$\int d^3k \left(-k \cdot k\tilde{G}(k) + k_0^2\tilde{G}(k) - 1 \right) e^{+ik\cdot x} = 0,$$

which only holds for all x, if the integrand vanishes, which follows directly by Fourier transforming both sides. Hence the Fourier transform of Green's function is

$$\tilde{G}(k) = \frac{1}{k_0^2 - k \cdot k}$$

and

$$G(x) = \frac{1}{(2\pi)^3} \int d^3k \frac{1}{k_0^2 - k \cdot k} e^{+ik\cdot x}. \qquad (2.5.5)$$

However, this integral can not be evaluated due to the denominator $k_0^2 - k \cdot k$ vanishing for $\|k\| = k_0$. By shifting the denominator off the real axis by an infinite amount and into the complex plane, the integral can be evaluated, however, the result depends on the direction of the shift. The validity of this approach is not discussed here, but the accompanying restrictions are well covered in the literature, see e.g. [63]. One obtains an advanced and a retarded Green's function given by

$$G_\pm(x) = \frac{1}{(2\pi)^3} \lim_{\epsilon \to 0} \int d^3k \frac{1}{(k_0 \pm i\epsilon)^2 - k \cdot k} e^{+ik\cdot x}.$$

The integral is evaluated by changing to spherical coordinates and using the Residual theorem in combination with Jordan's Lemma (see Appendix A.4). The limit can be taken and the Green's function is computed to

$$G_\pm(x) = -\frac{1}{4\pi} \frac{e^{\pm ik_0|x|}}{|x|}.$$

In the following, only the time-forward Green's function G_+ is considered, its difference compared to G_- is analog to the choice of a propagation direction, when obtaining the angular spectrum propagator in Section 2.3.2.1. It is worth noting, that the specific form of the Green's function as a spherical, outgoing wave is the basis of what is referred to as Huygens-Fresnel principle.

2.5.2 Born series

We now consider the inhomogeneous Helmholtz equation by separating the refractive index field $n(x)$ in a homogeneous part, equal to the vacuum refractive index

1, and its remaining difference η, which is referred to as scattering potential:

$$k_0^2 n^2(x) = k_0^2 + \eta(x)$$

It follows

$$\nabla^2 U(x) + k_0^2 U(x) = -\eta(x)U(x).$$

The general solution of the equation $U(x)$ can be separated into a field solving the homogeneous Helmholtz equation $U_0(x)$ and a field solving the inhomogeneous equation $U_S(x)$

$$U(x) = U_S(x) + U_0(x). \tag{2.5.6}$$

$U_0(x)$ can be considered to be the incident field as it would propagate, if no scatterer was present and thus $n(x) = 1$, i.e.

$$\nabla^2 U_0(x) + k_0^2 U_0(x) = 0.$$

$U_S(x)$ is the additional scattered field and it fulfills

$$\nabla^2 U_S(x) + k_0^2 U_S(x) = -\eta(x)U(x). \tag{2.5.7}$$

The scattered field can be written with the help of the Green's function

$$U_S(x) = -\int d^3x'\, G_+(x - x')\eta(x')U(x'), \tag{2.5.8}$$

which can be shown by inserting the expression (2.5.8) for U_S into (2.5.7). The problem with the obtained solution (2.5.8) is, that the field U_S appears on both sides of the equation. Inserting (2.5.6) in (2.5.8) gives

$$U_S(x) = -\int d^3x'\, G_+(x - x')\eta(x')U_0(x') - \int d^3x'\, G_+(x - x')\eta(x')U_S(x').$$

The expression (2.5.8) can be inserted recursively:

$$U_S(x) = -\int d^3x'\, G_+(x - x')\eta(x')U_0(x')$$
$$+ \int d^3x' \int d^3x''\, G_+(x - x')G_+(x' - x'')$$
$$\times \eta(x')\eta(x'')U_0(x')U_0(x'')$$
$$\mp \dots.$$

The resulting infinite series is known as the Born series. Breaking it after the first term, which is identical to replacing the total field U in (2.5.8) by the incident field U_0, is known as the first order Born approximation. In the first order Born approximation the total field can thus be written as

$$U(x) = U_0 - \int d^3x' \, G_+(x - x') \eta(x') U_0(x').$$ (2.5.9)

This means that the incident field is strong compared to the scattered field, and therefore scattering of already scattered light is neglected: The first order Born approximation only considers single scattered light, while higher orders take multiple scattering into account. Additionally, the effects of a constant refractive index, i.e. a change in the wavenumber of the propagating light can, within scattering theory, be seen as a wave propagating with the vacuum wavenumber, superimposed with a second scattered wave, effectively adjusting the wave field to the refractive index.

Let the incident wave be an ideal, monochromatic, plane wave, propagating in direction k_0 and given by

$$U_0(x) = A_I e^{+ik_0 \cdot x}.$$ (2.5.10)

The scattered field in first order Born approximation is then given by

$$U_S(x) = \frac{A_I}{4\pi} \int d^3x' \, \frac{e^{ik_0|x-x'|}}{|x-x'|} e^{ik_0 \cdot x'} \eta(x').$$

2.5.3 The Ewald's sphere

Emil Wolf first showed that optical holography in combination with the Born series can be used to effectively obtain Fourier components of the scattering potential η. By changing incident angle of the plane wave and the angle of the detection hologram, the complete Fourier plane can be obtained [64]. The following shows a slightly different derivation of the resulting optical scattering formula.

Inserting the Fourier representation of the Green's function (2.5.5) in the first order Born approximation (2.5.9) of the scattered field and still assuming an ideal plane incident wave as given by (2.5.10), one gets

$$U_S(x) = -\frac{A_I}{(2\pi)^3} \int d^3x' \int d^3k' \frac{1}{(k_0 + i\epsilon)^2 - k' \cdot k'} e^{ik' \cdot (x-x')} e^{ik_0 \cdot x'} \eta(x').$$

Exchanging the integrals and introducing the Fourier transform of the scattering potential $\bar{\eta}$ this gives

$$U_S(x) = -\frac{A_I}{(2\pi)^3} \int d^3k' \frac{1}{(k_0 + i\epsilon)^2 - k' \cdot k'} \cdot e^{ik' \cdot x} \cdot \bar{\eta}(k' - k_0). \qquad (2.5.11)$$

Inserting angular spectrum of the scattered field

$$U_S(x,y,z) = \frac{1}{(2\pi)^2} \int dk'_x \int dk'_y \, \tilde{U}_S\left(k'_x, k'_y, z\right) e^{+ik'_x x + ik'_y y}$$

into (2.5.11) and assuming that the identity needs to hold for all x and y yields that the integrands need to be identical:

$$\tilde{U}_S\left(k'_x, k'_y, z\right) = -\frac{A_I}{2\pi} \int dk'_z \frac{1}{(k_0 + i\epsilon)^2 - k'^2_x - k'^2_y - k'^2_z} e^{+ik'_z z}$$

$$\times \bar{\eta}\left(k'_x - k_{x0}, k'_y - k_{y0}, k'_z - k_{z0}\right).$$

By introducing $k^2_z = k^2_0 - k'^2_x - k'^2_y$, using the Residual theorem in combination with Jordan's Lemma for integration, and finally taking the limit $\epsilon \to 0$, the integral can be evaluated to

$$\tilde{U}_S\left(k'_x, k'_y, z\right) = -A_I \frac{i}{2k_z} e^{+ik_z z} \bar{\eta}\left(k'_x - k_{x0}, k'_y - k_{y0}, k_z - k_{z0}\right). \qquad (2.5.12)$$

As can be seen from (2.5.12), the angular spectrum of the scattering field $\tilde{U}_S$ is proportional to the angular spectrum of the scattering amplitude at spatial frequencies shifted by the wave vector of the incident field. For monochromatic light, the frequency components of the scattering amplitude that can be obtained this way are on a sphere called the Ewald's sphere shown in Figure 2.5.1a. Figure 2.5.1b illustrates, that by modifying the incident wave vector k_0 the sphere can be rotated about the origin which lies on the sphere. This way all frequency components of the Fourier representation of the scattering potential $\bar{\eta}(k)$ with $\|k\| < 2\|k_0\|$ can be obtained. After inverse Fourier transforming the scattering potential can thus be obtained with a uniform resolution of $\lambda/2$. We will investigate the relation of scattering theory and the Ewald's sphere to holoscopy in Section 5.4.4.

The acquisition and reconstruction of data using this technique is known as diffraction tomography and results have been shown [65–67]. Apart from electromagnetic waves [68], this technique has also been applied to acoustic (see e.g. [69,70]) and seismic (see e.g. [71]) waves.

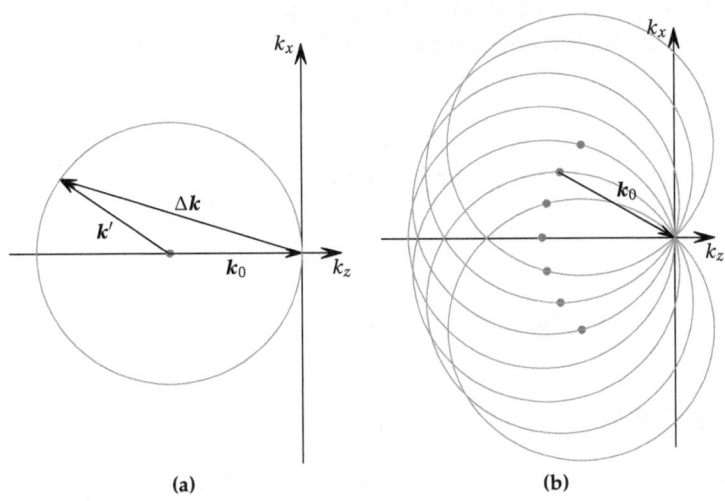

(a)   (b)

Figure 2.5.1.: The Ewald's sphere. a) Using a single incident wave vector k_0 and rotating the detector (camera) around the sample, a sphere of values in the Fourier space of the scattering potential can be obtained. b) By also rotating the incident wave vector, the sphere is rotated around the origin and all frequency values with $\|k\| < 2\|k_0\|$ can be obtained.

2.5.3.1 Diffraction grating

An additional result, that follows from the forward scattering formula (2.5.12), is the grating equation. By choosing the coordinate system appropriately, a grating can be a periodic structure with spatial frequency k_0 in the x-direction, an extended object in the y-direction and sharply localized in the z-direction. The Fourier transform of its scattering potential can therefore be written as

$$\tilde{\eta}(k_x, k_y, k_z) = \delta(k_x - k_0)\delta(k_y).$$

By using (2.5.12), for the incident and outgoing beam it follows immediately, that the first δ-distribution enforces

$$k'_x - k_{x0} = k_0,$$

where k'_x is the x-component of the scattered wave vector, k_{x0} is the x-component of the incident wave vector and k_0 is the spatial frequency of the grating. As the former two are proportional to the wavenumber, the vectorial change of direction due to diffraction at a grating is inversely proportional to the absolute wavenumber and proportional to the frequency of the grating.

By introducing ingoing and outgoing angles φ_i and φ_o, related to the k-components by

$$\sin \varphi_i = \frac{k_{x0}}{k}, \quad \sin \varphi_o = \frac{k'_x}{k},$$

it follows with $k = 2\pi/\lambda$ the grating equation

$$\frac{2\pi}{k_0}(\sin \varphi_o - \sin \varphi_i) = \lambda, \tag{2.5.13}$$

where $2\pi/k_0$ is the spacing of the grating.

2.6 Holography

With the propagator (2.3.6) introduced in Section 2.3.2.2, it is obvious, that once a monochromatic wave field is known in one plane, it can be computed in any other plane assuming free space. In fact, if the reflected wave of an object is fully captured in an out-of-focus layer, it can be propagated back (refocused) to its origin numerically. Classical imaging does not provide the full information on the wave field, but rather only captures its intensity (2.3.5), the squared amplitude of the wave. Although methods have been found to reduce the defocus of classically obtained images, for example by using blind deconvolution and regularization techniques, their complexity is increased and results are still not optimal. Other methods try to capture the complete light field by using specialized cameras [72,73], but for coherent imaging another, simpler possibility exists.

2.6.1 Classical holography

It was Dennis Gábor, who invented in the 1940s a method, called holography, that was capable of capturing not only the amplitude, but also the phase of a wave field. Although it was first used for electron microscopy [74], Gábor could demonstrate first optical in-line holograms using filtered light sources [75]. With the invention of the laser in 1962, Yuri Denisyuk and Emmett Leith together with Juris Upatnieks demonstrated first optical holograms of extended objects [76,77]. Finally, in 1971, Dennis Gábor won the Nobel Prize of Physics for "his invention and development of the holographic method" [78].

Holography is based on interference. The object wave $O(x)$ to be fully captured, including its phase and amplitude, is superimposed with a well reproducible reference wave $R(x)$ on a photographic plate. The resulting intensity distribution on the plate located in the plane $z = z_0$ can be described by

$$
\begin{aligned}
I(x,y,z_0) &= \gamma |O(x,y,z_0) + R(x,y,z_0)|^2 \\
&= \gamma \Big(|O|^2(x,y,z_0) + |R|^2(x,y,z_0) \\
&\quad + (RO^*)(x,y,z_0) + (R^*O)(x,y,z_0) \Big).
\end{aligned}
$$

If the photographic plate is developed, it will have a local transmittance $T(x,y)$, that is a linear function of the original intensity distribution, i.e. $T(x,y) = a - bI(x,y,z_0)$, with a and b depending on the photographic material and the exposure time. Illuminating the holographic plate with the exact same reference wave that was used to capture the hologram in the first place, the field in the hologram plane is proportional to the transmittance distribution multiplied with the reference wave:

$$
\begin{aligned}
R(x,y,z_0)T(x,y) = a - b\gamma \Big(\big(R|O|^2\big)(x,y,z_0) + \big(R|R|^2\big)(x,y,z_0) \\
+ \big(R^2O^*\big)(x,y,z_0) + \big(|R|^2O\big)(x,y,z_0) \Big). \quad (2.6.1)
\end{aligned}
$$

The respective field contains four terms:

- $R|O|^2$: DC and autocorrelation term of the object wave field. The spatial frequency range covered by this term, depends on the lateral resolution and is twice as large as the frequency range covered by the object field. Its center is located in the $(0,0)$-frequency in frequency space.

- $R|R|^2$: DC and autocorrelation term of the reference wave field. In most cases the reference wave field is a simple spherical or plane wave. This term therefore is very well localized in spatial frequencies. Its center is located in the $(0,0)$-frequency in frequency space.

- R^2O^*: Twin image term. It is proportional to the complex conjugated object wave. Due to the complex conjugation it is located on the opposite side of the $(0,0)$-frequency compared to the image term and will form a real image.

- $|R|^2O$: Image term. It is proportional to the original object wave and is in general the reproduction of the object wave field one tries to obtain, and a virtual image of the object. Using additional optical elements, a real image of the object can be created.

Thus the exact same object wave can be created, without the object being present anymore. The object wave can now be observed, as if the object was still there, either with the eye or any other imaging system. In Section 5.3.3.3 we will more closely investigate the frequency shifts and terms of the interference signal and also its wavelength dependency.

2.6.2 Digital holography

In 1977 Goodman et al. demonstrated that the holographic plate can be digitized and the image can be reconstructed sharply using the numerical diffraction demonstrated in Section 2.3.2.1 [79]. Contrary to the case of classical holography the reference wave does not need to be reproducible, but one rather requires its numerical representation – which is easily obtained for spherical or plane waves. For many applications the exact knowledge of the wave is actually not required. The reconstruction wave is even deliberately modified from the original reference wave, to shift the reconstructed images or introduce a magnification.

2.7 Optical coherence tomography (OCT)

2.7.1 Time-domain optical coherence tomography (TD-OCT)

In Section 2.3.4.1, it was shown that the amplitudes of the interference fringes in a Michelson interferometer decrease, if the path lengths in the two interferometer arms become unequal. How fast the amplitude drops with the displacement of the mirrors, depends on the coherence length of the light source. In time-domain optical coherence tomography (TD-OCT) this is used deliberately as an optical sectioning technique, the suppression of different path lengths is thereby referred to as coherence gate. Consequently, in TD-OCT, we replace the mirror in one of the two interferometer arms, further called sample or object arm, by a specimen composed of several layers, with the intention to analyze its depth structure. The backscattered wave at the photodiode is given as a superposition of incoherent waves reflected at various depths z' by

$$O(t) = \frac{A_O}{c} \int dz'\, \eta(z') \int d\omega\, e^{-i\omega t} A_C\left(\frac{\omega}{c}\right) e^{+i\frac{\omega}{c}(z_0 + 2z')},$$

where $\eta(z')$ denotes the backscattering potential, i.e. the amplitude of backscattered light relative to the amplitude of the incident light. Light attenuation in larger depths due to the backscattering or absorption are assumed small and are neglected, i.e. only the first order Born approximation is considered (Section 2.5).

If the reference mirror is displaced from z_0 by a fixed distance z, the reference wave field is

$$R(z,t) = \frac{A_R}{c} \int d\omega' \, e^{-i\omega' t} A_C\left(\frac{\omega'}{c}\right) e^{+i\frac{\omega'}{c}(z_0+2z)}.$$

The interference signal can now be computed to

$$\langle R^*(z,t)O(t)\rangle_t \propto \int dt \, \frac{A_R}{c} \int d\omega' \, e^{i\omega' t} A_C^*\left(\frac{\omega'}{c}\right) e^{-i\frac{\omega'}{c}(z_0+2z)}$$

$$\times \frac{A_O}{c} \int dz' \, \eta(z') \int d\omega \, e^{-i\omega t} A_C\left(\frac{\omega}{c}\right) e^{+i\frac{\omega}{c}(z_0+2z')}$$

$$= 2\pi \frac{A_R A_O}{c^2}$$

$$\times \int dz' \, \eta(z') \int d\omega \, |A_C|^2\left(\frac{\omega}{c}\right) e^{i\frac{\omega}{c}2(z'-z)}. \qquad (2.7.1)$$

The obtained signal is thus the Fourier transform of the intensity spectrum convolved with the backscattering potential, showing a high frequency modulation. The resulting cross-correlation term can also be expressed using the temporal degree of coherence (2.3.17) by

$$= 2\pi \frac{A_R A_O}{c^2} \int dz' \, \eta(z') \gamma\left(2\frac{z'-z}{c}\right).$$

Here it can be seen, that the smaller the coherence length of the light source, the better the depth discrimination. A simulated cross-correlation signal of a TD-OCT for a single mirror in the sample is shown in Figure 2.7.1 and illustrates the acquired signals of a sample.

In general the outlined procedure does not provide any lateral discrimination of the sample structures. Most commonly, a laterally localized (Gaussian) beam is used in the interferometer in a setup similar to the one shown in Figure 2.7.2. This way, the depth profile at a specific lateral position is obtained, which is in analogy to ultrasound imaging, known as an A-scan. By laterally scanning this Gaussian beam, a sectional image is obtained, known as B-scan. This was first shown in 1991 by Fujimoto's group. [3].

2.7.2 Fourier-domain optical coherence tomography (FD-OCT)

In 1995 Fercher et al. showed that OCT can also be performed by acquiring multiple interference signals of coherent light at different wavenumbers, while leaving the

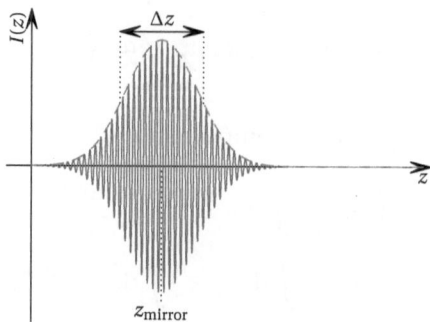

Figure 2.7.1.: Simulated TD-OCT interference signal $\mathrm{Re}\langle R^*O\rangle$ as a function of the reference mirror position. For the simulation, a mirror in the sample arm with $\eta(z) = \delta(z - z_{\mathrm{mirror}})$ was assumed.

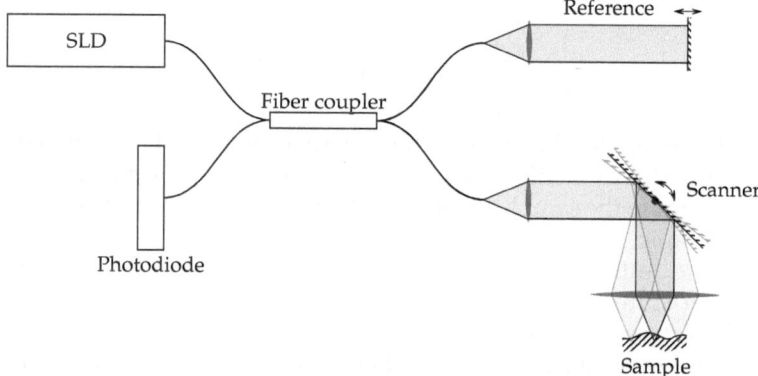

Figure 2.7.2.: Typical fiber based scanning time-domain OCT setup. The light from a fiber coupled broadband light source (superluminescent diode, SLD) is split by a fiber coupler in reference and sample and brought to an open beam confocal scanning mechanism (sample) and an open beam axial scanning reference. The scattered and reflected light is coupled back into the fiber and re-superimposed by the fiber coupler. The superimposed light is then measured by a photodiode.

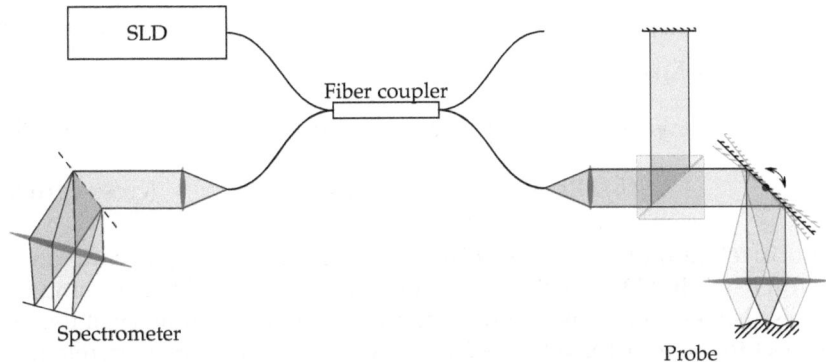

Figure 2.7.3.: Typical spectral-domain OCT setup. The light of a superluminescent diode is brought through a fiber coupler in a Michelson interferometer, where it is split by a beam splitter in reference and sample light. The sample light is scanned over the specimen, reference light is reflected by a mirror or a retroreflector. Finally, the light is re-superimposed by the beam splitter and brought onto the spectrometer via the fiber coupler.

reference arm fixed [80]. In this technique, known as Fourier-domain OCT (FD-OCT), the acquisition of the various signals is achieved either sequentially by using a rapidly tunable laser (swept-source) and a photodiode as first demonstrated by Chinn et al. [81], or by using a broadband light source and splitting it into its wavenumber components by using a diffractive element and a line camera, i.e. a spectrometer (spectral-domain) as shown by Bail et al. [9, 82]. All acquired wave fields can be superimposed numerically with a respective phase shift, introducing a numerical coherence gate, simulating a broadband light source after superposition. As shown below, this superposition is equivalent to performing a Fourier transform on the spectral data.

The coherent and monochromatic wave field of reference $R(k)$ and sample $O(k)$ are given by

$$R(k) = A_R A_C(k) e^{ikz_0},$$

and

$$O(k) = A_O A_C(k) \int dz'\, \eta(z') e^{ik(z_0 + 2z')},$$

where we now neglected any time-dependent signal terms as they cancel in the measurable coherent intensity signal (monochromatic light). The intensity signals obtained are given by

$$I(k) = \gamma |R(k) + O(k)|^2 \qquad (2.7.2)$$
$$= \gamma S(k) A_R^2$$
$$+ \gamma S(k) A_O^2 \int dz \int dz' \eta(z) \eta(z') e^{i2k(z-z')} \qquad \text{(autocorrelation)}$$
$$+ \gamma S(k) 2 \operatorname{Re} A_R A_O \int dz' \, \eta(z') e^{i2kz'}. \qquad \text{(cross-correlation)}$$

The autocorrelation term is caused by interference of the light scattered by the object with itself. In FD-OCT this causes noise and unwanted background images, sometimes also referred to as coherence noise. Due to the symmetry of the integral with respect to the transform $k \to -k$ it is a real signal. The cross-correlation term is the actual OCT signal, used to obtain the scattering potential $\eta(z)$. Performing an inverse Fourier transform on the cross-correlation term one obtains

$$\mathscr{F}_z^{-1}\left[S(k) A_R A_O \left(\int dz\, \eta(z) e^{-i2kz} + \int dz\, \eta(z) e^{+i2kz} \right) \right]$$
$$= A_R A_O \tilde{S}(z) * \left(\eta\left(-\frac{z}{2}\right) + \eta\left(+\frac{z}{2}\right) \right), \quad (2.7.3)$$

i.e. the scattering potential η can be reobtained, overlaid with its own mirror image, and convoluted with the Fourier transform of the intensity spectrum, defining the axial PSF and the resolution of the OCT system. However, if we assume that $\eta(z) = 0$ for all $z < 0$, i.e. there is no scattered light with an overall path length shorter than the reference path length, the most disturbing overlay with its own mirror image can be prevented; the real part of an analytic signal is acquired.

For many computations in FD-OCT it is totally sufficient to use the analytic signal belonging to a real acquired signal, especially if it is ensured during acquisition, that all path lengths of the sample light are longer than the path length of the reference light. By adding the Hilbert transform of the real acquired signal as imaginary part, the analytic signal can be obtained.

Just as TD-OCT, the principle of FD-OCT does not provide a lateral discrimination of the acquired structures; a laterally localized (Gaussian) beam is used and laterally shifted to scan the sample. A single depth scan, at a specific lateral position, is called A-scan. A sectional image, comprising many A-scans, is called B-scan.

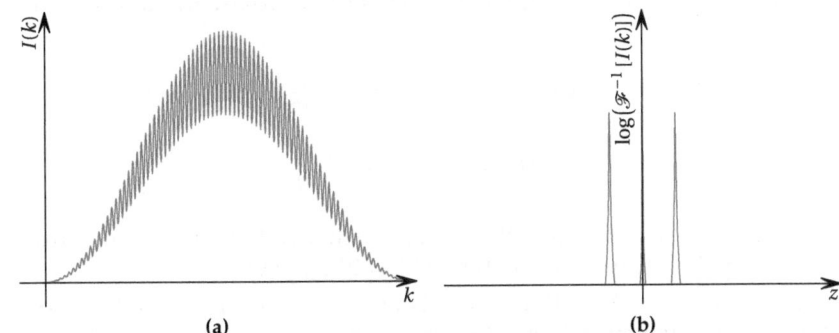

Figure 2.7.4.: a) Simulated signal of an intensity spectrum of a Fourier-domain OCT as a function of the wavenumber k, if a single mirror or reflecting surface is in the sample arm. b) Fourier transform of the simulated spectrum gives an A-scan. The symmetry with respect to the $z = 0$ axis is visible, as only the acquisition of the real-part of the cross-correlation term is possible.

2.7.2.1 The axial point spread function

As seen in (2.7.3), the reconstructed signal will be convolved with the Fourier transform of the spectrum $\tilde{S}(z)$. It should be noted that in general the spectrum will be centered at a certain wavenumber k_0 which gives a phase factor $\exp(ik_0z)$ after Fourier transform; this can be used as high precision measurement of the position z.

If the spectrum $S(k)$ is known precisely, the acquired autocorrelation term of (2.7.2) can be divided by $S(k)$ and multiplied with a more suitable spectrum, a window function $w(k)$. This spectral shaping, known as apodization, can significantly improve the axial point spread function of the signal. Various spectra and their Fourier transforms are shown in Figure B.1.1 and a quantitative comparison of the resulting axial intensity PSFs is shown in Figure B.1.2. In general, the smaller the signal width of the PSF, the more severe its side lobes and vice versa. As shown in Section B.1, assuming a window covers the complete input data of the DFT, the full width at half maximum (FWHM) of a rectangular window will be 1.21 pixels, whereas the FWHM of a Hann window will be 2 pixels.

2.7.2.2 Dispersion in FD-OCT

A dispersion mismatch of sample and reference arm in FD-OCT causes a degraded OCT signal. GVD mismatch between reference and sample arm introduces an additional path length difference $z_{\text{disp}}(k)$, that depends on the wavenumber k

(compare Section 2.3.3.1). The signal of the FD-OCT cross-correlation term (2.7.2) then reads

$$
\begin{aligned}
I_{\text{disp}}(k) &= \gamma 2 S(k) A_R A_O \, \text{Re} \int dz \, \eta(z) e^{i 2 k (z + z_{\text{disp}}(k))} \\
&= \gamma 2 S(k) A_R A_O \, \text{Re} \left(e^{i\phi(k)} \int dz \, \eta(z) e^{i 2 k z} \right).
\end{aligned}
\tag{2.7.4}
$$

These wavenumber dependent path lengths, further called dispersion, introduce a phase factor $\exp(i\phi(k))$ with a suitable $\phi(k) = 2 k z_{\text{disp}}(k)$.

The effect of $\phi(k)$ on the OCT signal is best analyzed by a Taylor expansion around the central wavenumber of the spectrum k_0:

$$
\phi(k) = \phi(k_0 + \Delta k) = \phi(k_0) + \left. \frac{\partial}{\partial k} \phi(k) \right|_{k=k_0} \Delta k + \mathcal{O}\left(\Delta k^2\right)
\tag{2.7.5}
$$

The constant term $\phi(k_0)$ will add an additional overall phase, while the term proportional to Δk will shift the complete image in z-direction by $\partial \phi / \partial k$. Higher order terms change the local frequencies on the spectrum and therefore lead to a broadening of the reconstructed OCT signal. Group velocity dispersion (GVD) therefore decreases axial resolution and causes a degradation in FD-OCT imaging quality.

2.7.2.3 Sensitivity advantage of FD-OCT

Although Fourier-domain OCT has a certain elegance to it, its final breakthrough only happened, when it was shown that it has a unique advantage compared to TD-OCT [82–85]: In case of a fixed irradiation of the sample and a fixed measurement time, FD-OCT can achieve a significantly higher sensitivity, compared to TD-OCT. The fundamental reason is the parallel acquisition of all depths in FD-OCT. By applying the inverse Fourier transform, each photon is assigned its scattering depth, whereas time-domain OCT gates and neglects the photons outside of the coherence gate and requires a repeated measurement for each depth.

2.7.2.4 Signal roll-off in FD-OCT

In FD-OCT a signal attenuation for higher measurement depths is observed, which is commonly referred to as signal roll-off. Higher depths are coded as higher frequencies in the OCT spectrum, and a limited spectral resolution causes this degradation of high-frequency signals.

More specifically, as shown in Section 2.2, the signal acquired in position space using rectangular shaped pixels (spectrometer-based FD-OCT) or in time by using a fixed integration time (swept-source OCT) will have a filter-effect that causes a signal degradation in frequency space. Additionally, the spectral resolution is limited due to the specific OCT technology applied. In spectrometer-based OCT it is the limited resolution of the spectrometer optics. In swept-source OCT it is the limited instantaneous coherence length of the laser, which is equivalent to a broadened line spectrum during the sweep.

2.7.3 Direct and heterodyne detection

2.7.3.1 Direct detection

When detecting an optical signal of intensity I the number of obtained photoelectrons will be proportional to the intensity, assuming the bandwidth of the signal is small, such that variations in photon energy $E = \hbar\omega$ and efficiency of the detector are neglectable. Assuming the number of photoelectrons is Poisson-distributed with the expectation value N_γ proportional to the intensity of the field and the σ-noise is $\sqrt{N_\gamma}$ (shot noise), the signal-to-noise ratio (SNR) is determined by

$$\text{SNR} = \frac{N_\gamma}{\sqrt{N_\gamma + N_e}}, \tag{2.7.6}$$

where N_e denotes additional intensity independent noise sources, such as electronic read-out noise, etc. If N_γ becomes small, N_e dominates the noise. If N_γ becomes large, the shot noise dominates the overall noise, the measurement is shot noise limited.

2.7.3.2 Heterodyne detection

In heterodyne detection the signal wave is encoded with a well-known reference wave, that is coherent to the sample wave; for example in OCT, the cross-correlation term of (2.7.2) is the actual detected signal. The relative number of photoelectrons in the cross-correlation signal is proportional to $2\sqrt{N_R N_O}$, where N_O and N_R are the number of object and reference photoelectrons, respectively. The complete photoelectrons reaching the detector are given by $N_R + N_O$, resulting in the shot noise $\sqrt{N_R + N_O}$. The SNR it thus given by

$$\text{SNR} = \frac{2\sqrt{N_R N_O}}{\sqrt{N_R + N_O + N_e}},$$

with N_e representing additional noise. In the case $N_R \approx N_O$, the signal-to-noise ratio is identical to the direct detection (2.7.6). However, a gain of the signal can be achieved by increasing the amount of reference light, in the limit $N_R \to \infty$, the SNR is given by

$$\text{SNR} = 2\sqrt{N_O},$$

and thus only determined by the number of photoelectrons from the sample arm. The read-out and electronic noise N_e becomes neglectable and the measurement is shot noise limited. Interferometric techniques, such as OCT can therefore detect single photons, even with a detector that has considerable readout noise.

2.7.4 Scanning OCT

The basic principles of time-domain and Fourier-domain OCT only provide methods for axial sectioning and thus they have to be combined with lateral imaging modalities. Most commonly, scanning OCT is used, possible setups are shown in Figure 2.7.2 for TD-OCT and Figure 2.7.3 for SD-OCT. An advantage of this approach is the confocal gating, which effectively rejects multiple scattered photons that would be assigned to a false depth otherwise.

2.7.4.1 Confocal gating

Confocal microscopy is widely used; it allows for very high-resolution imaging, and only detects structures in a certain layer of the sample. It works by using a Gaussian beam of high NA to illuminate a spot of the sample. The intensity will be maximal in the focus of the Gaussian beam, axially and laterally dropping rapidly a few Rayleigh lengths z_R and a few central beam widths w_0 away, respectively. By sharply imaging exactly this spot of the sample onto a pinhole, photons outside the region are gated. Parts of the sample that could provide larger amounts of light onto the pinhole have not been illuminated. To achieve this coherence gate, both the detection and the illumination is important. The detection principle is illustrated in Figure 2.7.5.

In strongly scattering samples, multiple scattered light in general no longer originates from the region that is imaged onto the pinhole. This light is rejected. Optical coherence tomography is usually combined with confocal imaging to suppress multiple scattered photons. In practice, a fiber can be used as pinhole.

A mathematical precise derivation of confocal imaging is rather complicated, and can be found in the literature [86–88].

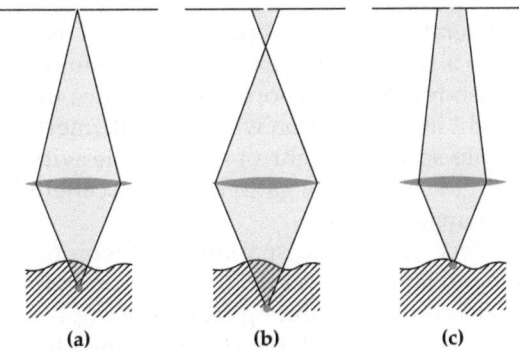

Figure 2.7.5.: Detection principle of a confocal gate: Only light that is imaged sharply onto a pinhole can be detected (a). Light from areas below (b) or above (c) the focus cannot pass the pinhole and is gated. In practice the illumination also plays a significant role but is not depicted here.

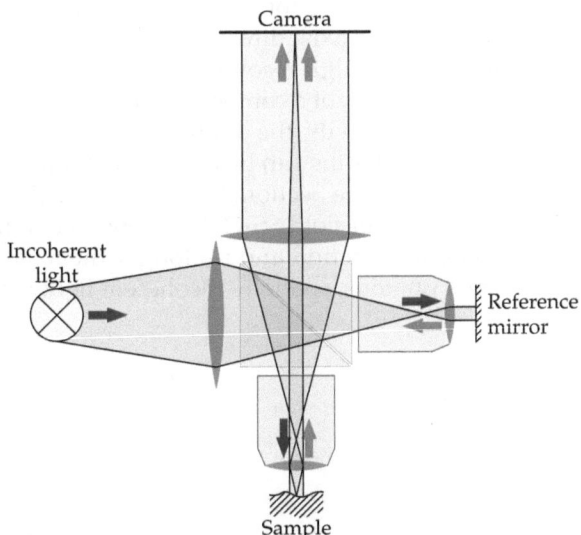

Figure 2.7.6.: Full-field time-domain OCT setup. A temporally and spatially incoherent light source is used and its light is split by a beam splitter in reference and sample arm. The light illuminates sample and reference mirror through identical microscope objectives. The backscattered and reflected light is imaged onto the camera with a suitable high magnification.

2.7.5 Full-field OCT

In full-field OCT the lateral acquisition is parallelized by using an area camera. A full-field TD-OCT setup is shown in Figure 2.7.6, it is commonly used for very high resolutions [23–26], also referred to as optical coherence microscope. It can use spatially incoherent light in combination with a Linnik interferometer to suppress interference of multiple scattered light of the sample with the reference light. This spatial coherence gate effectively gates multiple scattered photons, reducing artifacts and improves image quality.

The availability of very broadband spatially and temporally incoherent light sources, as e.g. a halogen lamp, allows axial resolution in the sub-micron range. With these light sources confocal scanning using fiber optics as in Figure 2.7.2 is impossible. Its advantage to confocal microscopy is mostly its higher sensitivity due to the heterodyne gain and its better depth discrimination. Its major disadvantage, compared to other OCT techniques, is the inefficient use of backscattered photons due to the lateral and axial gating. Consequently, the typical acquisition time is quite long.

Full-field Fourier-domain OCT has also been demonstrated [38, 39], a possible setup is shown in Figure 2.7.7. It allows for very high acquisition speeds because of the parallelization in both axes: in axial direction by FD-OCT, in lateral direction by the area camera. Going to very high resolutions is not feasible with full-field swept-source OCT. Although the lack of a confocal gating allows to collect photons from all depths with uniform sensitivity, the depth of focus decreases with higher lateral resolutions. Thus not all depths can be acquired at once. In this case, the sensitivity advantage of FD-OCT (see Section 2.7.2.3) is lost, axial scanning of the focus is required. Additionally, full-field SS-OCT is sensitive to motion artifacts as the fastest axis is the lateral direction and no longer the axial. To reduce the signals of multiple scattered photons, spatially incoherent light sources have been used [89].

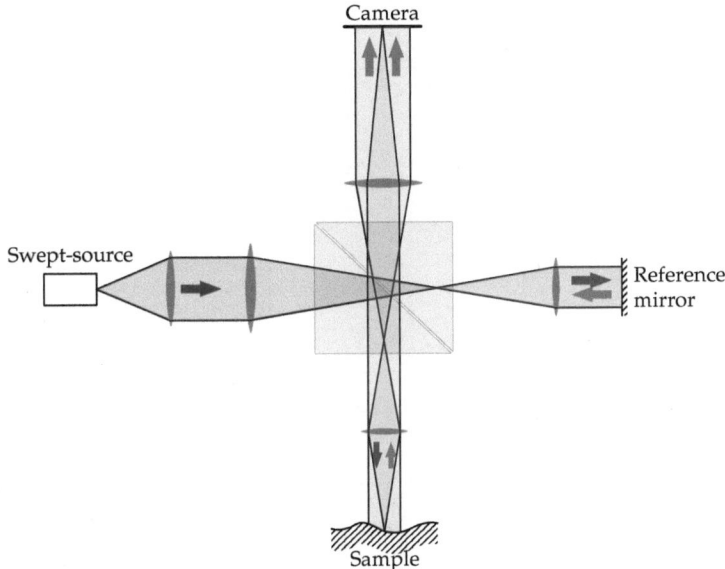

Figure 2.7.7.: Full-field swept-source OCT setup. Light of a tunable light source is split by a beam splitter in reference and sample arm. In the sample arm, the light illuminates the specimen, the backscattered light is imaged onto the camera, usually by applying an additional magnification. The reference light is collimated onto the camera by suitable imaging optics.

3

FD-OCT signal processing using the non-equispaced fast Fourier transform

In FD-OCT a Fourier transform needs to be performed to obtain depth information from the acquired spectral OCT data (Section 2.7.2). This Fourier transform needs to be performed with respect to the wavenumber k-axis, but most Fourier-domain OCT setups do not provide spectral data that is acquired as a function of the wavenumber k, and resampling needs to be performed prior to the actual Fourier transform. This resampling step is crucial to OCT imaging quality, and in this Chapter, different methods to perform a fast Fourier transform (FFT) on data points that are given on non-equispaced nodes will be evaluated.

This work was presented on the European Conference on Biomedical Optics (ECBO) 2009, Munich and this Chapter is based on the proceedings to the corresponding presentation [90].

3.1 The chirped FD-OCT signal

As shown in Section 2.7.2, Fourier-domain OCT uses the spectral dependence of the interference between light from a sample and a reference to determine the depth resolved reflectivity of the sample. Spectrometers usually do not record the spectrum as a function of the wavenumber k (see Section 2.5.3.1). For tunable light sources, as applied in SS-OCT, the wavenumber is in general not a linear function of time, however, sampling is often done with a fixed frequency. Consequently, in both scenarios, the spectral data is not acquired as a function of the wavenumber k, instead a chirped signal is obtained. Acquiring the OCT signal as a function of time or spectrometer-pixel, in the following commonly denoted as t, the FD-OCT cross-correlation term (compare (2.7.2)) is

$$I(t) = \gamma 2S(k(t))A_R A_O \mathrm{Re} \int \mathrm{d}z\, \eta(z) e^{\mathrm{i}2k(t)z}. \tag{3.1.1}$$

With a non-linear dependence of k on t, a Fourier transform with respect to the t-axis fails to invert this integral and results in broadened signals and significantly reduced imaging quality.

Wavenumber-linearized spectrometers [91] in SD-OCT or the syncing of the data acquisition of SS-OCT to a k-clock [92] can provide spectra which can be directly Fourier transformed. But wavenumber-linearized spectrometers fail to compensate the chirp completely, and in SS-OCT, jitter on the k-clock is difficult to avoid, which reduces performance of the OCT system and may introduce imaging artifacts.

The use of a non-linear-k data acquisition in combination with numerical compensation even offers advantages. It reduces the effects of disturbing aliased signals and electronic fixed frequency noise that cannot be filtered in spectrometer-based and full-field swept-source OCT. Finally, a software correction avoids increased technical complexity. The choice of algorithm for the resampling is important, because it determines resolution, side lobes of the axial PSF and signal-to-noise ratio.

3.2 Calibration

In order for any algorithm to correct the chirp $k(t)$ numerically, it needs to be quantified first. Apart from direct computation of the λ-to-k function from the wavelengths by taking the grating equation (Section 2.5.3.1) into account [93], several calibration methods exist. Computation can be combined with functional optimization [93], or, functional optimization can be used on its own [94]. A short-time Fourier transform can be used to directly compute the λ-to-k vector. In any way, the results of these methods offer almost indistinguishable results, providing an axial resolution at the theoretical optimum. A simple method is to use a filter, similar to the approach proposed in [95], that is closely related to determine the phase from the analytical signal. Assuming the spectral FD-OCT cross-correlation term $I(t)$ of a single reflecting surface, for example a mirror at depth z_0, has been obtained. It can be described by setting $\eta(z) = \delta(z - z_0)$ in (3.1.1):

$$I(t) = \gamma 2S(k(t))A_O A_R \cos(k(t)z_0). \tag{3.2.1}$$

Here, $k(t)$ is the wavenumber as a function of the sampling parameter t and is to be obtained. The acquired signal (3.2.1) is thus directly proportional to a cosine with fixed frequency in k-space. The cosine term can be expressed using Euler's relation $\exp(i\varphi) = \cos\varphi + i\sin\varphi$ by exponential functions

$$I(t) = \gamma 2A_O A_R S(k(t)) \frac{e^{+ik(t)z_0} + e^{-ik(t)z_0}}{2}.$$

By assuming that $dk(t)/dt > 0$ for all t (or $dk(t)/dt < 0$ for all t), the actual signal $I(t)$ can be interpreted as a sum of two well defined analytic signals – one with only positive frequencies, and one with only negative frequencies.

By filtering one of the two signals in the frequency domain (compare to the Hilbert transform in Section 2.1.1), for example removing all negative frequencies, the Fourier transform of the analytic signal

$$\tilde{I}_{\text{filtered}}(\omega) = \Theta(\omega) \cdot \tilde{I}(\omega)$$

is obtained, where ω denotes the Fourier conjugated variable corresponding to t and Θ is the step function (see (A.1.6)). Of the filtered signal

$$I_{\text{filtered}}(t) = \gamma 2A_O A_R S(k(t))e^{ik(t)z_0},$$

the argument is computed, which yields the phase modulo 2π

$$\arg I_{\text{filtered}}(t) = k(t)z_0 \bmod 2\pi.$$

If z_0 is chosen sufficiently below the Nyquist frequency, simple phase unwrapping methods can be applied. For larger z_0, zero padding of $\tilde{I}(\omega)$ prior to the inverse Fourier transform can increase the effective Nyquist frequency. The value of z_0 is not required in practice, if it is assumed that $k(t)$ has known domain in t and k space.

3.2.1 Calibration in presence of GVD mismatch

If the OCT cross-correlation term is additionally subjected to dispersion mismatch between reference and sample arm, the chirped FD-OCT cross-correlation term is given by combining (3.1.1) with (2.7.4) to

$$I(t) = \gamma 2S(k(t))A_R A_O \text{Re} \int dz\, \eta(z) e^{i2k(t)z + i\phi(k(t))}.$$

For static dispersion the phase factors (2.7.5) can be determined by using calibration measurements [95]: In analogy to the previous Section, one assumes two analytic signals of mirrors at positions $z = z_1$ and $z = z_2$ have been obtained, both subjected to the same dispersion $\phi(t)$ and the same chirp $k(t)$. These signals are given by

$$I_{\text{disp}}^{(1)}(t) = \gamma 2S(k(t))A_O A_R \exp\left(i\,\underbrace{(k(t)z_1 + \phi(k(t)))}_{f_1(t)}\right) \qquad (3.2.2)$$

and

$$I_{\text{disp}}^{(2)}(t) = \gamma 2S(k(t))A_O A_R \exp\left(i\,\underbrace{(k(t)z_2 + \phi(k(t)))}_{f_2(t)}\right). \qquad (3.2.3)$$

f_1 and f_2 shall denote the arguments of $I_{\text{disp}}^{(1)}$ and $I_{\text{disp}}^{(2)}$ after phase unwrapping. Using simple algebra this yields an expression for the chirp

$$k(t) = \frac{f_1(t) - f_2(t)}{z_1 - z_2}.$$

Contrary to the method shown in the previous Section, this method also works in the presence of a dispersion phase factor, but requires two calibration measurements. Again, the positions of z_1 and z_2 do not need to be known, if the domain of $k(t)$ is known in t and k space.

Using the function $k(t)$ the dispersion can now be determined for example by

$$\phi(t) = f_1(t) - k(t)z_1.$$

As shown in Section 2.7.2.2, the dispersion terms linear in k are in general not of interest in OCT signal restoration. Therefore the position of z_1 does not need to be known: one can develop the function $\phi(t)$ in polynomials of $k(t)$ and remove zeroth and first order. The zeroth order is merely a constant phase factor and the first order corresponds to a refractive index mismatch between sample and reference arm, which is usually compensated by adjusting the reference length to place the sample within the imaging window.

The previously stated calibration method requires two measurements of a strong reflection, e.g. a mirror, in two different positions z_1 and z_2. Especially for full-field

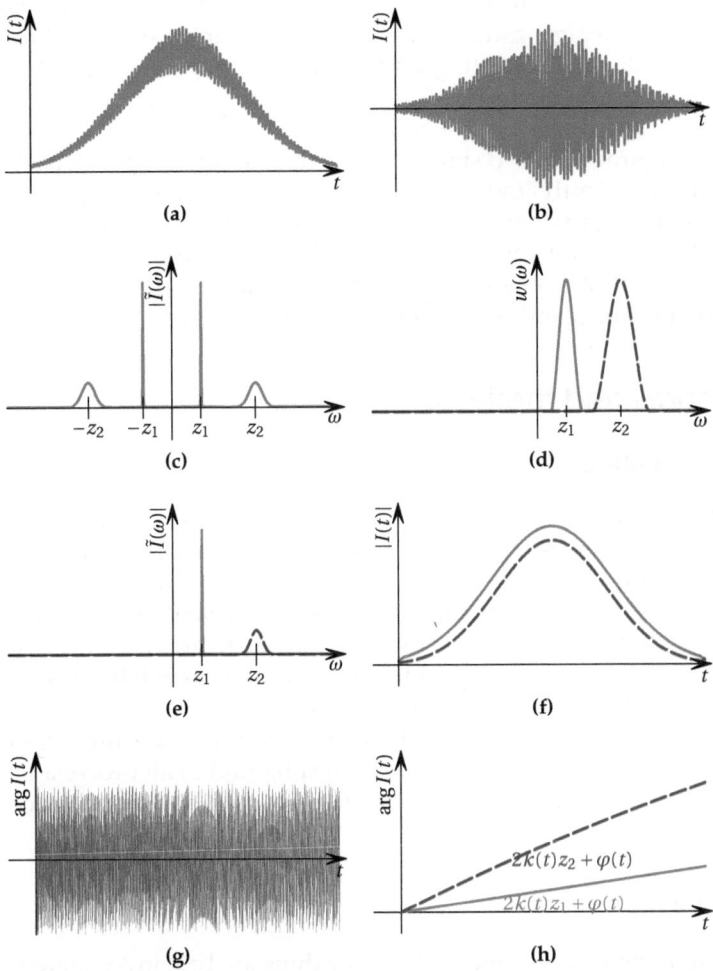

Figure 3.2.1.: Calibration of dispersion phase factors and chirp. a) Simulated spectrum as it would be acquired with signals from two depths. b) Spectrum after removing the DC part of the signal. c) A-scan with both arms; the symmetry of the signal is visible. d) Windows on one side of the A-scan that would entirely filter the two distinct signals. e) The two filtered A-scans, only having positive frequencies. f) Absolute value of the inverse Fourier transform of the two complex A-scans. g) Argument of the inverse Fourier transform of the two complex A-scans. h) Phase-unwrapping of the data shown in (g).

FD-OCT and – as will be shown in Section 5.6.2 – holoscopy this is quite cumbersome as quite large amounts of data need to be acquired for each measurement.

Instead of measuring a sample in two depths, a single extended object can be used, with two distinct reflecting surfaces at positions $z = z_1$ and $z = z_2$ – for example the reflecting surfaces of a cover slip. In principle the signal used for calibration can also be part of the acquired measurement that is to be corrected. As shown in Figure 3.2.1, two signal peaks can be filtered separately, if dispersion and chirp are sufficiently low and the two signals do not overlap. Two bandpass filters, centered around z_1 and z_2, are applied independently to the original signal with a bandwidth small enough that the other peak does not disturb the signal. Two analytic signals are obtained – completely analogous to (3.2.2) and (3.2.3); the determination of chirp and dispersion works accordingly.

3.3 Materials and methods

3.3.1 The algorithms

The algorithms to be compared are the linear interpolation and FFT with an oversampling α (iFFT α) as introduced in Section 2.1.3.2 and the non-equispaced FFT with oversampling α and cut-off parameter m (NFFT α, m) as introduced in Section 2.1.3.3 [58, 59]. The mathematical precise treatment, the NDFT, which was introduced in Section 2.1.3.1, has also been used. It provides the precise development into chirped sines and cosines. A similar approach has previously been suggested [96], which also takes dispersion into account.

To test these algorithms and to evaluate their performance, measured and simulated data were used. The signals were then subjected to all processing steps, that are usually used to obtain final OCT images. Imaging quality and reconstruction speed are compared.

3.3.2 Simulation

Degradation of the A-scans due to the algorithms are best investigated with simulated noise-less signals: the precision of the result is limited by numerical accuracy. The simulated signals were created with an artificial spectrum based on a Hann window shape (A.1.8) and they include five distinct, equidistant modulation frequencies and an artificial chirp given by

$$k(t) = t + \left[1 - \left(\frac{t - N/2}{N/2}\right)^2\right] \cdot h, \qquad (3.3.1)$$

with $h = 50$ and N being the number of pixels of the simulated spectrum. A quadratic chirp of this magnitude is usually encountered by a spectrometer-based OCT with 900 nm central wavelength and 150 nm bandwidth. The spectra were discretely convolved with a Gaussian function with $\sigma = 0.5$ pixel, to simulate an artificial cross-talk between sensor pixels, and a finite resolution of the spectrometer optics (spectral point spread function). The convolution results in an artificial roll-off of the signal intensity, similar to that observed for real devices (compare Section 2.7.2.4). Furthermore a spectrum with no chirp was created, where no interpolation and no oversampling was required. The resulting A-scans could be used as ideal reference.

3.3.3 Measured data

Real data were created from a highly reflecting surface in a Michelson interferometer, and OCT images of an infrared viewing card and of a contact lens were acquired. The highly reflecting surfaces of the test objects produced strong signals and artifacts from the different processing routines were visible well. Results of three different OCT devices with setups similar to the one shown in Figure 2.7.3 were compared. The devices had different imaging speeds and signal-to-noise ratios (SNR):

1. An OCT with an SNR of ~85 dB was very sensitive to the various processing techniques as small changes can be seen easily. For this purpose a Callisto from Thorlabs HL AG (Lübeck, Germany) was used, which has a full well capacity of $\sim 8 \times 10^6$ electrons and a sampling rate of ~1.2 kHz A-scan rate with 1024 pixels per spectrum

2. An ultra-high-speed OCT device with a significantly lower SNR of ~61 dB: the Hyperion from Thorlabs HL AG (Lübeck, Germany) uses a fast CMOS camera (Sprint SpL4096 70km, Basler, Ahrensburg, Germany) with an acquisition speed of ~127 kHz A-scan rate at 2048 pixels per spectrum. The full-well capacity of this device is 19,000 electrons.

3. A special version of the ultra-high-speed OCT device Hyperion, equipped with a linear-k-spectrometer, allowed OCT imaging without resampling. A prism in the spectrometer made the acquisition of data almost linear in k as demonstrated by Rollins et al. [91].

For all three devices, identical SLDs (SLD471-HP1-DIL-SM-PD, Superlum, Ireland) with central wavelength of 930 nm and FWHM of about 98 nm were used; the total spectral width imaged by the spectrometer was about 160 nm.

3.3.4 Signal processing

A complete signal processing chain, as used in FD-OCT, was performed for each acquired or simulated spectrum. The steps, to calculate the complete A-scans, starting from digital camera output, were the following:

1. casting the camera output after quantization (8-bit (Hyperion) or 16-bit (Callisto) unsigned integer data) to single precision (32-bit) floating point numbers.

2. removing offset errors by subtracting the dark signals of the camera.

3. apodization of the spectrum, i.e. dividing by a signal-less reference spectrum and multiplication with a suitable window function.

4. calculation of the A-scans by either FFT, NDFT, iFFT, or NFFT.

5. calculation of the logarithm of the absolute values of the complex output.

6. removing the signal roll-off caused by the interpolation.

All calculations were performed on a Core2 Quad with 2.6 GHz. The algorithms were implemented in C++ using the Intel C++ Compiler. To compute the FFT the FFTW library [57] was used. Multi-threading and vectorization using Intel Streaming SIMD Extensions (SSE) were used to optimize performance of the reconstruction. The measured times only include the processing, but not the acquisition time of the camera. A set of camera data was acquired and stored in RAM. This also removed any additional load for data acquisition from the central processing unit (CPU). Afterwards it was repeatedly processed to get an average measure for performance of the different algorithms. The resulting times hence include full processing from input data to final OCT image data.

3.4 Results and discussion

3.4.1 Processing speed

The processing throughput was measured for the discrete Fourier transform on non-equispaced nodes (NDFT), the fast Fourier transform (FFT), the linearly interpolated fast Fourier Transform (iFFT) and the fast Fourier transform on non-equispaced nodes (NFFT) with different oversampling and cut-off parameters. The performance of the NFFT algorithms depended slightly on the image data, namely on the function $k(t)$, which determines how many data points need to be evaluated for the pre-FFT convolution. The other algorithms did not depend on $k(t)$.

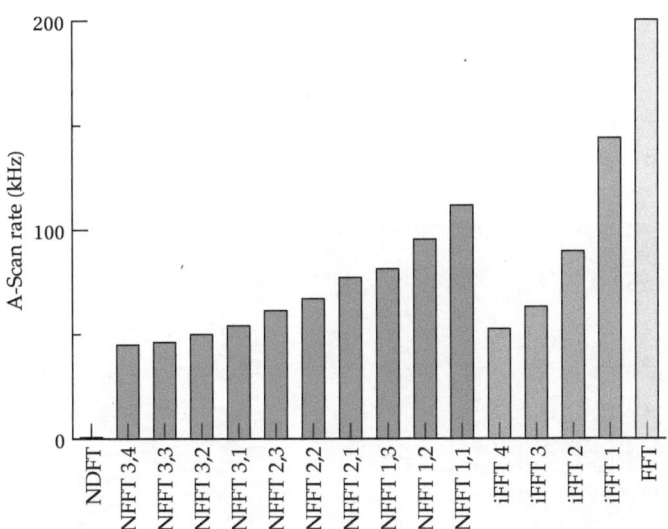

Figure 3.4.1.: Benchmark results of various processing routines for FD-OCT signal processing.

For the data measured with the linear-k-spectrometer Hyperion with 2048 pixels, the FFT achieved a throughput of more than 180 kHz A-scan rate (Figure 3.4.1). The NDFT, which needs a full matrix multiplication, could only calculate 600 A-scans per second, making it about 300× slower compared to the simple FFT. Linear interpolation without oversampling reduced the throughput to 140 kHz A-scan rate for the interpolated FFT (iFFT 1) and to 110 kHz A-scan rate for the NFFT 1,1. The speed reduction is caused by the additional interpolation. Oversampling by a factor of α further decreased the throughput roughly by a factor of $1/\alpha$. For the iFFT only two original data points are used for computing an interpolated value, whereas for the NFFT this number might vary slightly as it is specified as cut-off distance in k-space. For this reason the, NFFT 1,1 shows slightly worse performance than the iFFT1. The NFFT performance drops with increased α or m. With the computer hardware used, online processing of Hyperion data, acquired at 127 kHz A-scan rate, was only possible with the FFT or the iFFT 1.

3.4.2 Image quality with simulated data

Reconstruction quality was first analyzed with simulated data of five reflecting layers in different depths (Figure 3.4.2). The spectra were distorted by a typical chirp for spectrometer-based OCT according to (3.3.1). All noise and artifacts

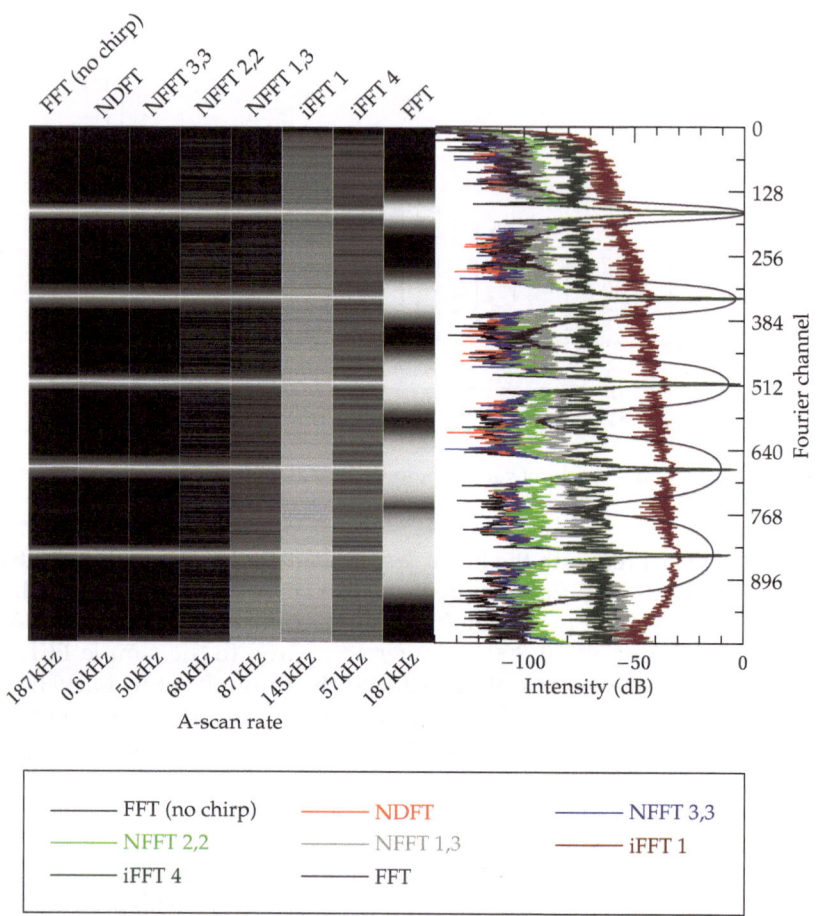

Figure 3.4.2.: Processing of simulated chirped OCT data without noise. The simulated chirp was typical for a spectrometer-based OCT with 900 nm central wavelength and 150 nm bandwidth. As reference the FFT of simulated unchirped data is shown left. At the right side all A-scans are plotted in one diagram for quantitative comparison in one diagram. The results are mostly distinguished by their noise floor.

observed in the processed data have to be attributed to numerical artifacts, since no photon or detector noise was added to the data. The iFFT 1 and the NFFT 1,1 (data not shown) introduced significant additional numerical noise which is more than 60 dB above the numerical noise of the NDFT or of a FFT on unchirped data. The simple FFT on the chirped data did not introduce additional noise, but suffered from unacceptable broadening of the structures.

Both, increasing the oversampling or the cut-off parameter reduced the numerical noise significantly. For the iFFT 4 and the NFFT 1,3 the noise is reduced by 30 dB to 40 dB and the NFFT 2,2 and NFFT 3,3 reached nearly the noise level of the NDFT. However, the processing speed of the NFFT 3,3 was nearly 100 times faster compared to the NDFT. Compared to the iFFT 4, the NFFT 2,2 achieved a lower noise at a slightly higher throughput (68 kHz vs. 57 kHz). The non-chirped input signal subjected to a simple FFT created the best possible result. For the chirped input data, the faster iFFT or NFFT are, the more numerical artifacts they produce. NFFT in general shows significantly less numerical noise and artifacts than the iFFT at similar processing speed. However, one should keep in mind that for real world situations the data precision is in general not limited by numerical accuracy, but by shot noise and detector noise. Therefore, depending on the SNR of the OCT device at hand, the numerical artifacts may be irrelevant, as they are covered by real OCT noise.

3.4.3 Image quality with measured data

It was expected that the superior performance of the NFFT should also be visible in high-SNR OCT A-scans, recorded with the Callisto. Images of reflecting surfaces, which were recorded at three different depths and with three different signal strengths, from 47 dB to 99 dB, were processed by the iFFT 1, the iFFT 4, the NFFT 2,2 and the NDFT, which served as reference for the optimal image quality (Figure 3.4.3). Increased noise levels and artifacts due to the processing became more visible as the signal strength was increased and the object was located at higher depths. For the interpolated FFT without oversampling (iFFT 1) images at 47 dB look virtually similar to the NDFT images. For higher signal strengths a significantly increased noise level was observed. Thus the artifacts in processing of real images will depend not only on the OCT device (maximum SNR) but also on the object (depth and reflectivity). This was improved with four time oversampling (iFFT 4), the images had only a slightly increased noise level at 99 dB signal strength compared to the NDFT. The NFFT 2,2, while still being faster (compare with Figure 3.4.1), gave even better images that can hardly be distinguished from the NDFT reference image.

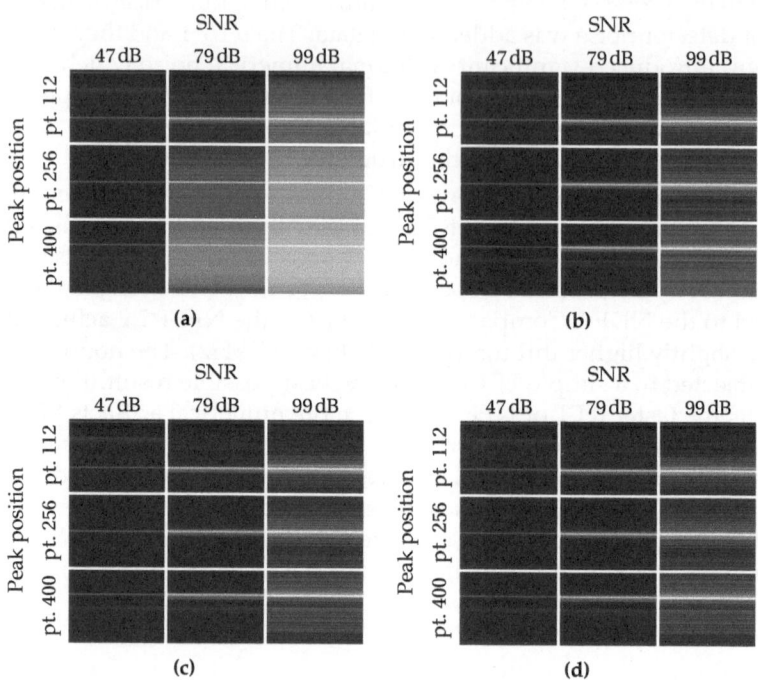

Figure 3.4.3.: Influence of depth and SNR on the imaging artifacts. Images of a reflecting surface in three different depths were recorded at three different signal levels. OCT images were recorded by a Thorlabs Callisto OCT device and processed by the iFFT 1 (a), the iFFT 4 (b), and the NFFT 2,2 (c). For comparison images were also processed by the NDFT (d), which is expected to give optimal, nearly artifact-free images. The system noise level was approximately 25 dB for all images, the increase in noise floor is due to numerical artifacts.

For a more systematical comparison, the performance of the different algorithms was compared at maximal signal of the Callisto in three different depths (Figure 3.4.4). The images show some artifacts, common to all processing routines, which are caused by the OCT itself. When the object was placed in the upper part, multiple equidistant lines appeared, that were caused by harmonics of the modulation frequency due to detector non-linearity. A white band at the end of the measurement range in Figure 3.4.4b appeared with all processing routines and was therefore probably not caused by data evaluation, but by the device itself. The iFFT algorithms improved significantly with increased oversampling α, but even for $\alpha = 4$ one can see artifacts reducing the image quality (see Figure 3.4.4c) compared

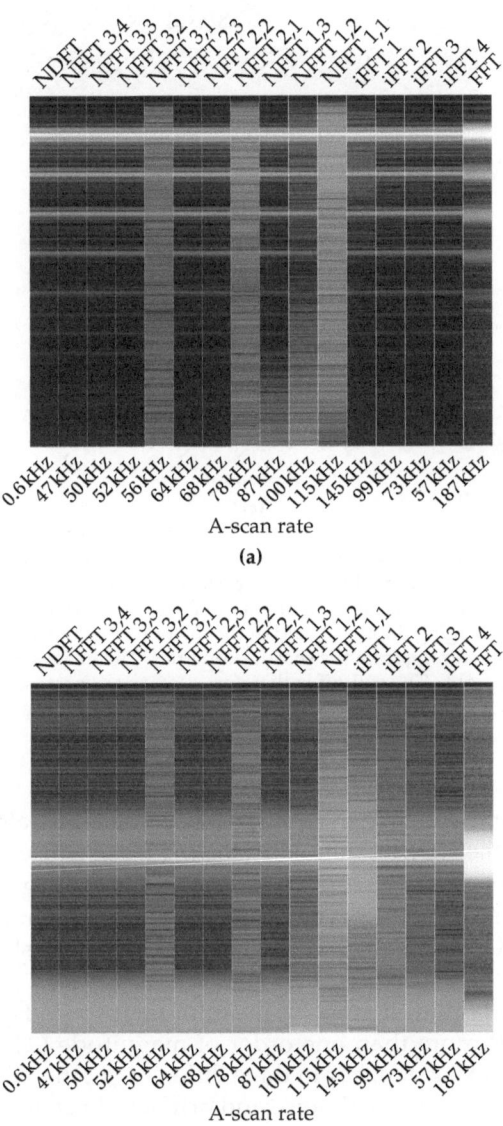

A-scan rate

(a)

A-scan rate

(b)

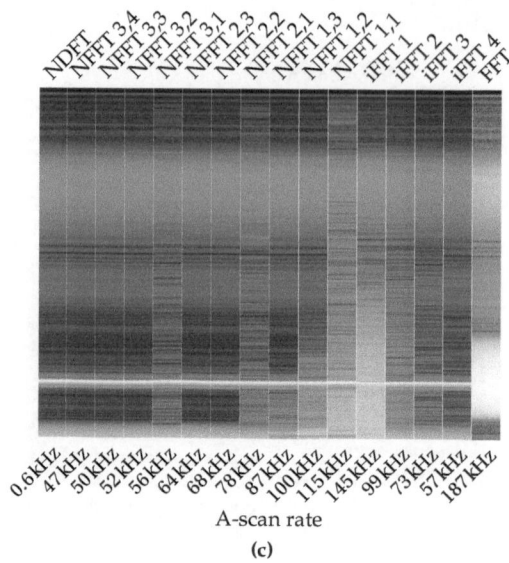

A-scan rate

(c)

Figure 3.4.4.: Results for the different processing algorithms when used with a Thorlabs Callisto OCT device ($8 \times 10^6\,e^-$ FWC, 85 dB SNR). For each image a single high peak was created using a Michelson interferometer at different depths, in the upper part of the imaging depth (a), in the middle of the imaging depth (b), and in the lower part of the imaging depth (c). Other signals are artifacts coming from the sensor properties. The images were normalized to the maximum peak height. The roll-off is therefore seen as an increased noise level. For the different processing routines, different noise levels are observed that are caused by numerical artifacts.

to the ideal result of the NDFT. In comparison, some NFFT algorithms cannot be distinguished from the NDFT, while still showing faster processing than the iFFT 4.

Finally the OCT image of an infrared card is shown in Figure 3.4.5. The iFFT 4 shows good results here, however artifacts are slightly reduced when using NFFT 2,2 instead. Using NDFT instead of NFFT does not show any further improvements.

Due to the reduced full well capacity, the ultra-high-speed Hyperion OCT device has a maximal SNR more than one order of magnitude (24 dB) lower than the Callisto. Measurements of a mirror in three different depths, similar to Figure 3.4.3, show greatly reduced numerical noise and artifacts, because of the higher noise level (Figure 3.4.6). The NFFT 1,3 and the iFFT 3 are virtually indistinguishable from the "exact" NDFT. In the upper part of the field of view and for signals with low modulation frequencies even the iFFT 1 yields good results while still being

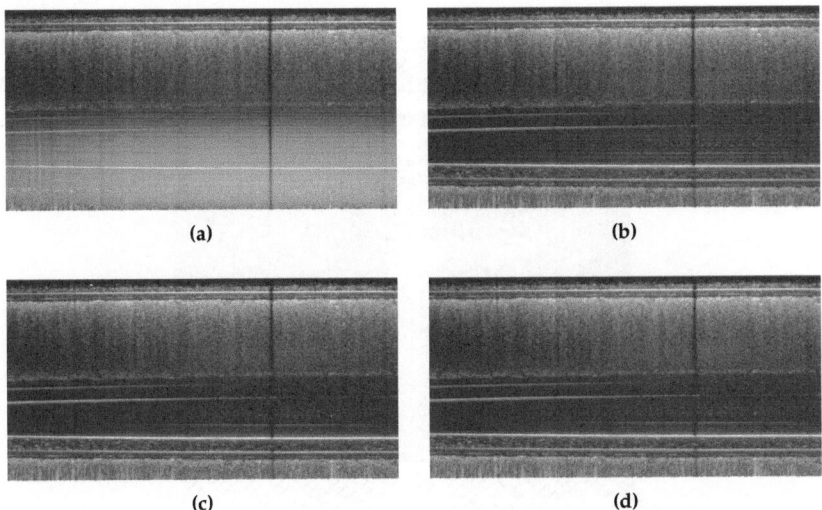

Figure 3.4.5.: OCT image of an infrared viewing card, acquired with a Callisto device for various processing routines. a) Using the iFFT 1 major artifacts are observed. b) Using the iFFT 4 the artifacts are reduced significantly, but still visible. c) Using the NFFT 2,2 no artifacts are observed, when compared to (d). d) Results of the NDFT as gold standard.

fast enough for online processing of the data ($\sim$150 kHz processing rate). Only for strong signals and high modulation frequencies (Figure 3.4.6c) increased side lobes can be seen with this algorithm. Compared to the Callisto, the Hyperion showed an increased fixed pattern noise of the OCT device itself, which is visible as vertical line in all processing routines. When the Hyperion OCT devise is used with a linear-k-spectrometer, the A-scans can be calculated with a direct FFT, which gives the highest processing speed (Figure 3.4.7). In the upper part of the depth range the images are even better compared to NFFT 1,1, NFFT 2,1, and NFFT 3,1. However, at the lower part of the measurement range, resolution is degraded due to a residual non-linearity of the spectrometer. When this is compensated by iFFT or NFFT, the best results were seen. However, online processing was only possible with the standard FFT and iFFT 1, which both have either noticeable broadened signals and slightly decreased SNR or show artifacts and side lobes.

Images of both Hyperion devices were also made of a contact lens which has several strong reflections (Figure 3.4.8). The artifacts of an iFFT 1 seen in Figure 3.4.6 appear also in the image of the contact lens (Figure 3.4.8b and 3.4.8d) while they are no longer seen for NFFT 2,2 (Figure 3.4.8e and 3.4.8c). Such strong reflections

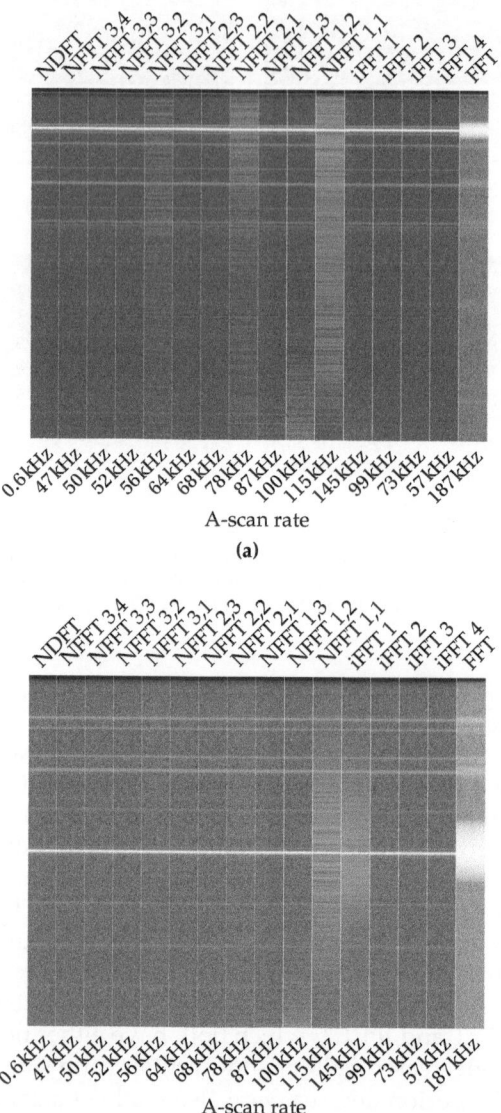

A-scan rate

(a)

A-scan rate

(b)

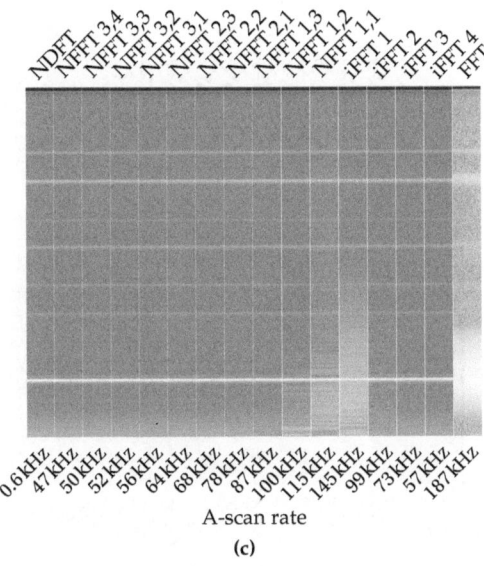

(c)

Figure 3.4.6.: Results for the different processing algorithms when used with a Thorlabs Hyperion OCT device (19,000 e^- FWC, 61 dB SNR). A single high peak was created using a Michelson interferometer at different depths, in the upper part of the imaging depth (a), in the middle of the imaging depth (b), and in the lower part of the imaging depth (c). Other signals are artifacts, originating form the camera properties in combination with the applied modulation. The images were normalized to the maximum peak height and were colorized in an area of 61 dB. The roll-off is therefore seen as an increased noise level.

are seldom for scattering probes and can be removed by using a suitable algorithm. When using the linear-k-spectrometer in real-world scenarios, residual chirp causes minor image degradation when using a standard FFT (Figure 3.4.8a), which is not entirely compensated by an iFFT 1 (Figure 3.4.8b). Additionally, imaging artifacts caused by the camera (Figure 3.4.8a, 3.4.8b, and 3.4.8c) can be easily mistaken for real reflecting surfaces, whereas these are strongly blurred after compensation of the images acquired by the normal spectrometer (Figure 3.4.8d and 3.4.8e).

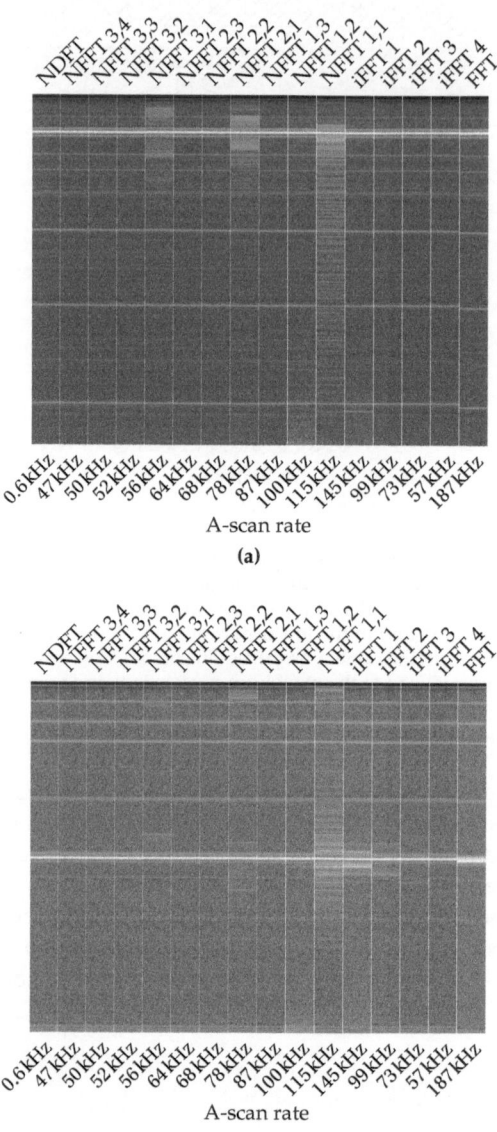

(a)

(b)

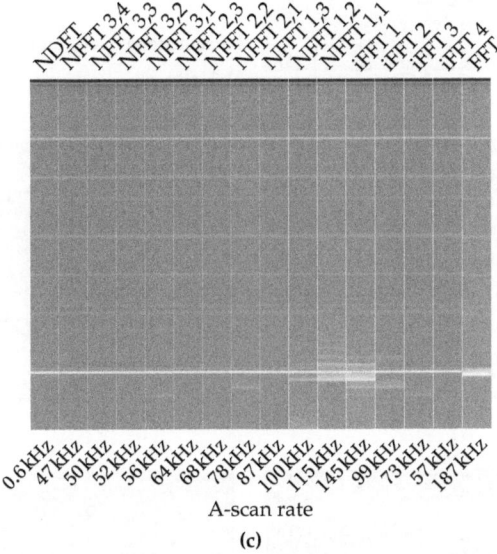

A-scan rate

(c)

Figure 3.4.7.: Results for the different processing algorithms when used with the Thorlabs Hyperion OCT device with linear-k-spectrometer (19,000 e^- FWC, 61 dB SNR). For each image a single high peak was created using a Michelson interferometer in different depths, in the upper part of the imaging depth (a), in the middle of the imaging depth (b), and in the lower part of the imaging depth (c). Other signals are artifacts, originating form the camera properties in combination with the applied modulation. The images were normalized to the maximum peak height and were colorized in an area of 61 dB. The roll-off is therefore seen as an increased noise level.

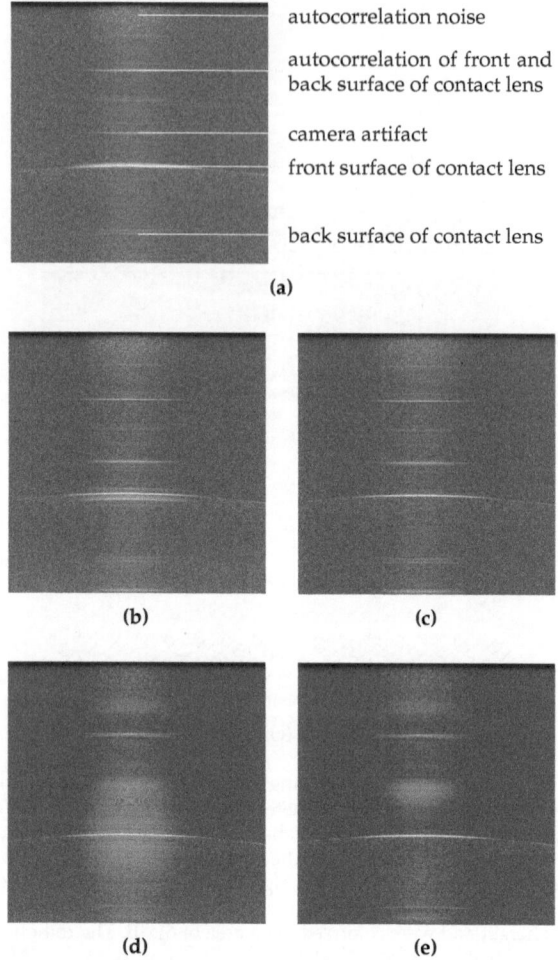

Figure 3.4.8.: OCT image of a contact lens, acquired with a Hyperion device, with and without linear-k-spectrometer. For computation time performance see also Figure. 3.4.1. a) Results using a standard FFT with a linear-k-spectrometer. b) Results using the iFFT 1 with a linear-k-spectrometer. The processing could be performed in real-time. c) Results using the NFFT 2,2 with a linear-k-spectrometer. d) Results using the iFFT 1 with a standard non-linear-k-spectrometer. The processing could be performed in real-time. e) Results using the NFFT 2,2 with a standard non-linear-k-spectrometer.

4 Motion and dispersion correction in FD-OCT

Sample motion in OCT imaging causes loss of signal and, in the case of swept-source OCT, loss of resolution. Especially in full-field swept-source OCT, sample motion during the volume acquisition is a problem, because of the parallel acquisition of up to a million A-scans, the time of a single sweep is dramatically increased. Artifacts are caused by axial and lateral motion, as well as rotations of the sample. But axial motion plays a special role, as its effect is increased by the Doppler shift. It is, in its effect and mathematical treatment, comparable to group velocity dispersion (GVD) mismatch between reference and sample arm in FD-OCT. Both result in axial blurring of the images. In this Chapter, a common approach to detect and correct both, axial motion and GVD mismatch, will be derived and demonstrated.

This Chapter is based on a publication in Optics Express, entitled "Common approach for compensation of axial motion artifacts in swept-source OCT and dispersion in Fourier-domain OCT", that we published in 2012 [97].

4.1 Introduction

Full-field swept-source (FF-SS) OCT (see Sections 2.7.5) is a way to increase imaging speed by massively parallel acquisition of A-scans. Compared to scanning OCT, the A-scan acquisition time of FF-SS-OCT increases dramatically. It becomes identical to the acquisition time of a complete volume and instead of digitizing the photocur-

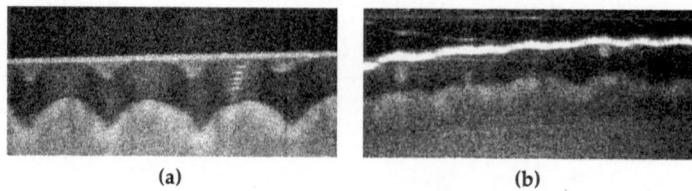

(a)   (b)

Figure 4.1.1.: B-scans of a finger tip, obtained by full-field swept-source OCT with different
        stabilization and imaging speed. a) Volume scan was acquired in ∼1 s, corresponding to
        1,000 fps. The finger was stabilized by pushing it against a glass plate. No motion artifacts are
        visible. b) Volume scan of the finger tip, acquired in ∼30 ms, which corresponds to 36,000 fps.
        The finger was stabilized by pushing it against a ring. Axial blurring of the image due to
        sample motion is visible.

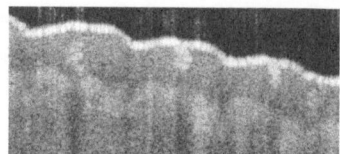

Figure 4.1.2.: B-scan obtained by FD-OCT of a finger tip with dispersion mismatch between
        reference and sample arm. Axial blurring of the image is clearly visible.

rent of a single pixel, a complete camera image is acquired for each wavenumber.
FF-SS-OCT has successfully been demonstrated to work *in vitro* [38,89] and also
for *in vivo* measurements of the human retina [39], but the requirements on the
acquisition speed of the digital camera are enormous to prevent motion artifacts.
A high-speed CMOS camera is needed, acquiring data with up to 100,000 fps or
more. At these acquisition rates only data of a greatly reduced area of interest are
recorded, which cannot be transfered continuously to a computer. As demonstrated
in Figure 4.1.1, motion artifacts remain a major problem for *in vivo* measurements
of not stabilized tissue. During the resulting sweep the acquisition time of each
single frame is fast enough to record the fringe pattern without significant loss of
contrast, but the sample motion distorts the phase relation between the different
frames as the wavenumber is scanned.

4.1.1 Doppler effect on axial motion in swept-source OCT

In general, sample motion is composed of rotations, axial, and transverse move-
ments. Due to the Doppler effect, axial motion has an additional impact on the
imaging artifacts. According to Section 2.7.2, the cross-correlation part of an inter-

ference signal with a mirror in the sample arm at a time-dependent position $z(t)$ is obtained by setting $\eta(z') = \delta(z' - z(t))$ in (2.7.2). This yields

$$I(t) = \gamma 2S(k)A_R A_O \mathrm{Re}\exp(\mathrm{i}2kz(t)),$$

with t being the acquisition time. In swept-source OCT the wavenumber is time-encoded and one can introduce a sweep curve $k(t) = m_k t + k_i$, with m_k being the sweep rate, k_i being the initial wavenumber at $t = 0$, $t \in [0; T]$, and $k_f = k(T)$ the final wavenumber. If the mirror is moving with constant velocity v, its position can be described by $z(t) = z_0 + vt$ and the interference signal becomes

$$I(t) = \gamma 2S(k(t))A_R A_O \mathrm{Re}\exp(\mathrm{i}2(m_k t + k_i)(z_0 + vt)).$$

Resorting all terms and rewriting the sweep rate as $m_k = \left(k_f - k_i\right)/T$ and the velocity as $v = \Delta z/T$, where Δz is the total physical displacement over the sweep time, the signal becomes

$$I(t) = \gamma 2S(k(t))A_R A_O\, \mathrm{Re}\exp\left[\mathrm{i}2\left(m_k t\left(z_0 + \frac{k_i}{k_f - k_i}\Delta z\right) + m_k vt^2 + k_i z_0\right)\right].$$

The depth information is extracted from the term proportional to t in the argument, the term quadratic in t leads to a broadening of the signal, and the t-independent term will merely add a phase. But the actual physical shift Δz is increased by a factor $k_i/\left(k_f - k_i\right)$. On these conditions, the Doppler effect increases the impact of axial motion by a factor of approximately $k_i/\left(k_f - k_i\right) = \lambda_f/\left(\lambda_i - \lambda_f\right)$, where λ_i and λ_f denote the initial and final wavelengths of the sweep, respectively [98]. This effect, which magnifies the axial motion by typically one order of magnitude, renders the axial direction the most vulnerable axis in a full-field SS-OCT system and causes axial blurring in the resulting image. For example, for a system with a center wavelength of 860 nm and a sweep range of 50 nm, the axial direction is approximately 17 times more sensitive to motion artifacts than the transverse directions. The following treatment is therefore restricted to axial motion.

4.1.2 Dispersion in FD-OCT

The axial motion causes a chirp of the momentary frequency of the interference as the path length continuously changes. This introduces additional time-dependent phase factors to the complex interference signal. These phase factors are comparable to the effect of unmatched group velocity dispersion (GVD) between sample and reference arm, the latter is usually caused by different optical components, but can also result from the sample itself. Its effect is demonstrated in Figure 4.1.2, where axial blurring is clearly visibly. More importantly, for high-resolution FD-OCT its correction is crucial, but only matching of the interferometer arms does often not yield optimal results [99–101]. As an example, the length of the human ocular bulb varies individually by 20 % [102]. For retinal imaging, it introduces GVD and thereby influences resolution and only an individual correction of GVD yields optimal image quality. In order to correct the mismatch, it needs to be determined correctly and afterwards GVD correction is possible using optical means [103], but this is expensive and cumbersome. Alternatively, dispersion correction is easily done numerically by a simple multiplication of phase factors prior to the fast Fourier transform (FFT).

4.2 Effect and correction of sample motion and GVD mismatch on the OCT signal

The complex representation (analytical signal) of the cross-correlation term of an FD-OCT interference signal $I(k)$ is given according to (2.7.2) by

$$I(k) = 2\gamma S(k) A_R A_O \int dz\, \eta(z) \exp(-i2kz). \tag{4.2.1}$$

For simplicity, the (conjugated) analytical signal will be used (Section 2.1.1). Axial movement of the sample during the wavenumber sweep, or GVD mismatch between reference and sample arm, both introduce an additional path length difference $z_{disp}(k)$, that depends on the wavenumber k (see also Section 2.3.3.1):

$$
\begin{aligned}
I_{disp}(k) &= 2\gamma S(k) A_R A_O \int dz\, \eta(z) e^{-i2k(z+z_{disp}(k))} \\
&= 2\gamma S(k) A_R A_O e^{-i\phi(k)} \int dz\, \eta(z) e^{-i2kz}
\end{aligned}
\tag{4.2.2}
$$

These wavenumber dependent path lengths, further called dispersion, introduce a phase factor $\exp(-i\phi(k))$ with $\phi(k) = 2kz_{disp}(k)$. Once $\phi(k)$ is known, its effect can simply be reverted by multiplying $I_{disp}(k)$ in (4.2.2) with the complex conjugated phase term $[\exp(-i\phi(k))]^* = \exp(+i\phi(k))$ prior to calculating the A-scans by the inverse Fourier transform.

The effect of $\phi(k)$ on the OCT signal is best analyzed by a Taylor expansion (see also (4.2.3)) around a central wavenumber k_0:

$$\phi(k) = \phi(k_0 + \Delta k) = \phi(k_0) + \left.\frac{\partial}{\partial k}\phi(k)\right|_{k=k_0} \Delta k + \mathcal{O}\left(\Delta k^2\right) \qquad (4.2.3)$$

Only considering zeroth and first order in the interference signal (4.2.2), the recon-structed A-scan obtained after the Fourier transform is

$$\eta_{\text{rec}}(z) \quad \propto \quad \mathscr{F}_z^{-1}[S(k)] * \mathscr{F}_z^{-1}\left[\int dz\, \eta(z)\, e^{-i2kz - i\phi(k_0) - i\partial_k\phi(k)|_{k=k_0}(k-k_0)}\right]$$

$$= \quad \mathscr{F}_z^{-1}[S(k)] * e^{-i\left(\phi(k_0) + \partial_k\phi(k)|_{k=k_0} k_0\right)} \eta\left(\frac{z}{2} + \partial_k\phi(k)|_{k=k_0}\right), \quad (4.2.4)$$

where $*$ denotes the convolution operation and $\mathscr{F}_z^{-1}[S(k)]$ is the PSF of the OCT signal. The constant phase term $\phi(k_0)$ only adds an additional constant phase. The linear term $\partial_k\phi(k)|_{k=k_0}\Delta k$ shifts the complete image due to (A.3.3). Higher order terms cause axial blurring of the reconstructed signal η_{rec}.

4.3 Determination of the correcting phase function

4.3.1 Cross-correlation of sub-bandwidth reconstructions

The in Section 3.2 described algorithms for determining sampling chirp and dis-persion require a reference measurement with a strongly reflecting structure and therefore are not applicable for correcting motion or sample induced GVD. An alternative approach is the cross-correlation of sub-bandwidth reconstructions (CCSBR), an algorithm that can retrieve the correcting phase function $\phi(k)$ from acquired raw data of any sample without specific calibration measurement. Since the sampling chirp in spectrometer-based FD-OCT is constant and can easily be determined by a reference measurement, we assume that data are acquired as a function of the wavenumber k. The algorithm can easily be adapted using a predefined chirp function $k(t)$ (see Chapter 3).

The basic idea for retrieving the phase term $\phi(k)$ from the image data is to apply a shifted windowing function $w(k)$ to the dispersion affected spectral data $I_{\text{disp}}(k)$ prior to the Fourier transform. This windowing function has a k-bandwidth small enough that the linear approximation of $\phi(k)$ in (4.2.4) becomes valid as demonstrated in Figure 4.3.1. The center of the windowing function k_0' is then shifted along the whole spectrum. The resulting short-time Fourier transform

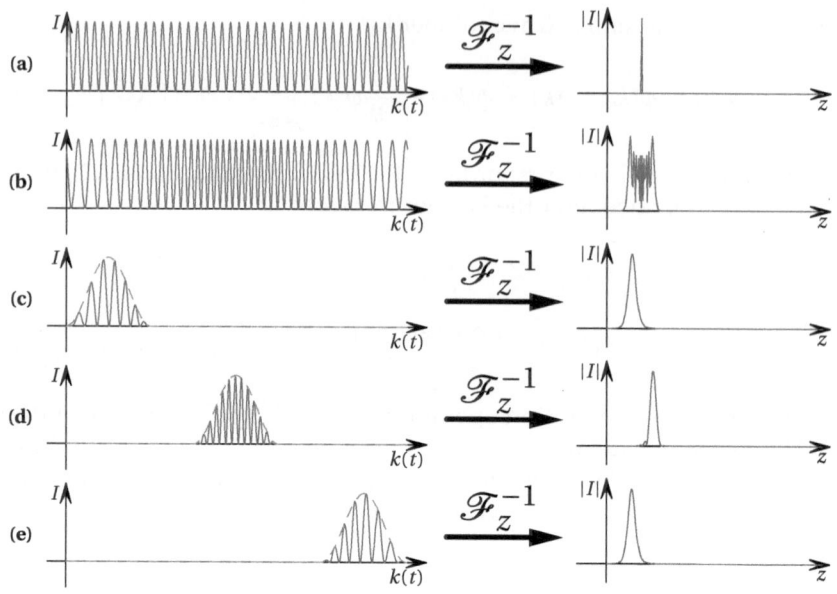

Figure 4.3.1.: Influence of dispersion on an FD-OCT signal and its short-time Fourier transforms. a) Spectrum of the interference from one reflecting surface in the sample. The width of the reconstructed OCT-Signal (point spread function) is determined by the spectral width. b) By dispersion, here visible by a chirp in the sine, the peak in the Fourier transform is broadened and modulated. c)-e) By filtering a small region of the spectrum in which the chirp is neglectable, shifted peaks are restored. However, these peaks are broadened compared to the unchirped case due to the reduced spectral bandwidth, introduced by the windowing.

(STFT) is given by

$$
\begin{aligned}
\mathrm{STFT}_{k_0'}\Big[I_{\mathrm{disp}}(k)\Big] &= \mathscr{F}^{-1}\Big[w(k-k_0')\,I_{\mathrm{disp}}(k)\Big] \\
&\propto \mathscr{F}^{-1}\Big[w(k-k_0')S(k)\int \mathrm{d}z\,\eta(z)\mathrm{e}^{-\mathrm{i}2kz-\mathrm{i}\phi(k)}\Big] \\
&= \mathrm{e}^{-\mathrm{i}\big(\phi(k_0')+\partial_k\phi(k)|_{k=k_0}k_0'\big)} \\
&\quad\times\Big\{\Big(\mathrm{e}^{-\mathrm{i}k_0'z}\mathscr{F}^{-1}\big[w(k)S(k)\big]\Big) \\
&\quad\quad *\eta\Big(\frac{z}{2}+\partial_k\phi(k)|_{k=k_0'}\Big)\Big\}.
\end{aligned}
\tag{4.3.1}
$$

With dispersion the apparent depth of the image structures depends on the local frequencies, which depend on k_0' as seen in Figure 4.3.1. From the displacement of the structures calculated by the STFT at different k_0' the phase $\phi(k)$ can be extracted. For reflections at interfaces this can be done by simple peak detection. For more complicated samples a cross-correlation of the signal intensities at the different k_0' with the STFT at a fixed wavenumber k_0'', e.g. the center wavenumber of the OCT source, can determine the shift. The cross-correlation can be effectively implemented by using fast Fourier transforms (FFTs) and the displacement $\Delta z(k_0') = \partial_k \phi(k)|_{k=k_0'}$ can be determined by finding the maximum of the cross-correlation

$$
\begin{aligned}
\partial_k \phi(k)|_{k=k_0'} &\approx \Delta z(k_0') \\
&= \arg\max_z \left(\left| \mathrm{STFT}_{k_0'} \left[I_{\mathrm{disp}}(k) \right] \right| \star \left| \mathrm{STFT}_{k_0''} \left[I_{\mathrm{disp}}(k) \right] \right| \right)(z),
\end{aligned}
$$

where $\arg\max_z$ denotes the z-value at the maximum of the argument and $\star$ denotes the cross-correlation. This procedure to obtain the derivative of the phases is illustrated in Figure 4.3.2. The required phase function $\phi(k)$ can then be calculated down to a linear term by the integration of $\Delta z(k)$

$$
\phi(k) \approx \int \mathrm{d}k\, \Delta z(k) + C. \tag{4.3.2}
$$

The arbitrary integration constant C defines a constant phase factor of the A-scans which has no influence on the resulting image intensity. The choice of k_0'' determines the linear term of $\phi(k)$ and therefore the axial position at which the dispersion corrected image will appear. In the case of object motion or GVD mismatch this is usually not the exact physical position. Also the apparent displacement $\Delta z(k)$ is not identical to physical motion of the sample. The cross-correlation of sub-bandwidth reconstruction (CCSBR) of the recorded images leads directly to a correction phase function which can eliminate the dispersion related image blurring.

The proposed algorithm is limited by a trade-off between the accuracy and spectral resolution with which $\Delta z(k_0') = \partial_k \phi(k)|_{k=k_0'}$ can be determined. A large variance of the window function σ_k^2 of the STFT increases the depth resolution of the displacement, but decreases the spectral resolution by which dispersion or motion is calculated. The uncertainty relation states, that the standard deviation of a signal cannot be arbitrary small in time and frequency at the same time, i.e. $\sigma_z \cdot \sigma_k \gtrsim 1/2$ (compare Appendix A.3.3). The displacement can however be determined more accurately than predicted by the uncertainty relation: σ_z^2 determines only the variance of the signal peak (axial PSF) of the cross-correlation. The position of the

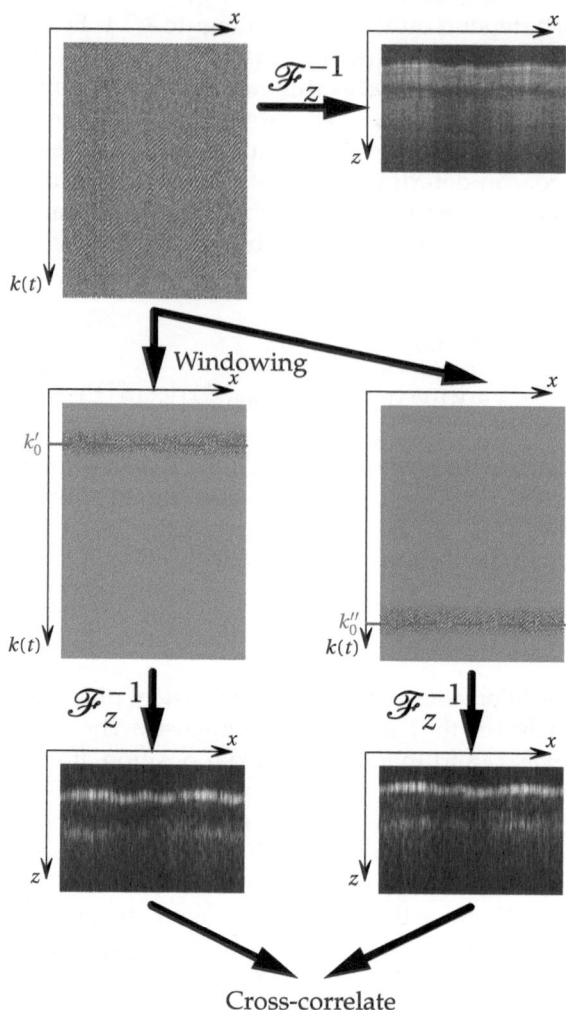

Figure 4.3.2.: Schematic representation of the algorithm, determining the dispersion phase function $\phi(k)$ by cross-correlation of sub-bandwidth reconstructions (CCSBR). The spectral data are windowed at different center wavenumbers k_0' and k_0'' and Fourier transformed to obtain depth information. The resulting OCT images will be shifted for each k_0' and k_0'' along the z-axis due to non-constant motion or GVD mismatch. Cross-correlating of these OCT images gives their apparent z-shift and by integration of the z-shifts the phase function is obtained.

maximum can be determined more precisely by means of zero-padding prior to the STFT and local fits to the maximum (see Section 4.4.1).

Other methods to correct for sample dispersion and motion blur rely on functional optimization techniques by iteratively changing coefficients of polynomial approximations of the dispersion in order to optimize axial resolution of the OCT data at hand [104]. These techniques will fail for sample motion induced phase factors, if the sample motion cannot be described by a polynomial of sufficiently low degree. Complexity and computation time increase significantly with each additional degree of freedom while robustness decreases. Additionally, optimization algorithms rely on correct axial resolution determination of the data and need a stopping criterion. Hence, there is need for a non-iterative algorithm, which can determine the phase error in SS-OCT data which is introduced by axial sample motion or GVD mismatch in the interferometer arms.

In the following Sections it will be shown, that the cross-correlation of sub-bandwidth reconstructed images can retrieve the phase errors directly from the acquired OCT data and it does not require any additional calibration measurements. It only uses the spectral raw data that were subjected to axial motion or dispersion, respectively. No image quality determination algorithms or optimization techniques are required. It is demonstrated, that the algorithm can determine the phase errors in full-field SS-OCT data caused by sample motion in *in vivo* measurements and can correct dispersion of the ocular media in retinal imaging.

4.4 Materials and methods

The new CCSBR algorithm was tested for motion correction with a full-field SS-OCT and for GVD mismatch correction in two different spectrometer-based OCT devices.

4.4.1 Implementation

Preprocessing Often FD-OCT signals contain fixed pattern noise from the detection electronics or from self-interference of reflecting surfaces in the optical setup. Both induce image structures with fixed positions in the B-scans or volume data. These imaging artifacts are independent of dispersion and sample motion. If sufficiently strong, they obscure the detection of the axial shifts.

For spectrometer-based FD-OCT fixed pattern artifacts are effectively removed by dividing the spectral data by a modulation-free reference spectrum and apodization of the data to a suitable window function, e.g. a Hann window. The phase in SS-OCT data is not stable enough for this approach. Removing such artifacts in the final intensity B-scans by subtracting a reference image destroys the phase information that is required for a successful phase detection. Subtracting these patterns in the STFT data destroys large parts of the image due to the reduced spatial resolution in the STFT data and a (negative) fixed pattern is introduced to the images.

Instead, the fixed pattern noise was efficiently removed by lateral high-pass filtering. A similar approach was successfully used as lateral filtering when creating complex spectral data in full-range OCT [105]. Here its only purpose is to remove low-frequency components. The measured spectra of a complete 3D-volume or a series of B-scans were Fourier transformed with respect to k and x, y directions. Fixed patterns show up in the respective depth as bright spots close to the lateral zero frequencies of the Fourier transformed spectra. High-pass filtering with respect to the x and y directions effectively removes the fixed patterns (see Figure 4.4.1), while maintaining most of the image information including the phases, which are located due to the speckle noise at high frequency regions.

Detection algorithm   In our experiments a Hann window was used for the STFT, which effectively reduces side lobes. However, the Hann window also reduces the spatial resolution (FWHM) z_{FWHM} compared to a rectangular window (Appendix B.1). Good results have been achieved using a window length between 64 and 128 pixel for spectra with 1024 data points. Outside the window all data points were set to zero. After Fourier transforming with respect to the k-axis, this zero-padding provides effectively an interpolation of the data. Of the Fourier transformed data sets, only the first 513 pixel were considered, i.e. only the positive frequency part of the FFT. The absolute values of these data points were used for cross-correlation. For an increase of the signal-to-noise ratio (SNR) the cross-correlation can be restricted to regions of interest which contain image information with high contrast and good SNR.

The cross-correlation was effectively implemented using FFTs, which calculate a circular cross-correlation if no additional zero padding is applied. However, as long as the Δz-shifts remain smaller than half of the measurement depth, the displacement can still be determined uniquely. As the effect of lateral shift was neglected in the motion detection scenario and as no lateral shift is present in the GVD mismatch case, the robustness of the algorithm can be increased by only searching the maximum of the cross-correlation for zero lateral shifts, i.e. $\Delta x = 0$ and $\Delta y = 0$. For sub-pixel accuracy of the displacement detection, a parabola going through the logarithm of the maximum intensity value and the two neighboring data points were computed. The argument of the maximum of the parabola was taken as the displacement between the cross-correlated sub-bandwidth reconstructions. In general it was not required to compute the displacement and the STFT

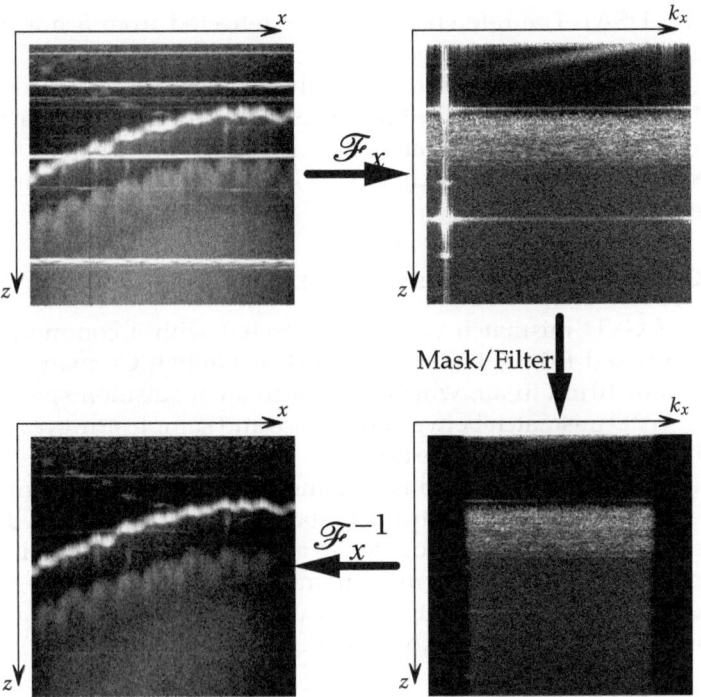

Figure 4.4.1.: Fixed pattern removal by filtering the A-scan series. From the B-scans or volume (upper left) the lateral FFT of the analytic A-scans is computed (upper right). Then a mask/high-pass filter is applied (lower right) and the inverse lateral Fourier transform gives the artifact-free image (lower left).

for every possible window position k'_0. A spline interpolation was used to get the intermediate data points and the resulting data was numerically integrated by building partial sums in order to calculate $\phi(k)$ from (4.3.2).

4.4.2 Full-field FD-OCT setup for axial motion experiments

The setup for full-field SS-OCT, which is described in more details in [39], used a tunable laser (Superlum Broadsweeper BS840-1) with a tuning range from 873.5 nm to 823.5 nm, which output was split in reference and sample arm (Figure 2.7.7). The light in the sample arm illuminated the object with a collimated beam, and the backscattered and reflected light was imaged onto either one of two monochrome high-speed CMOS cameras (A503k, Basler AG, Germany and FASTCAM SA5,

Photron Inc., USA). The reference light was reflected from a mirror and then collimated onto the camera. Camera and light source were synchronized using a start trigger of the tunable laser to start the acquisition of 1024 images by the camera. Maximal frame rate of the Photron SA5 camera depended on the number of pixels used for imaging. With an image size of 512×192 pixel a frame rate of up to 36,000 frames/s was reached and all 1024 images were acquired in a total acquisition time of 28 ms.

4.4.3 FD-OCT setup for GVD mismatch experiments

Correction of GVD mismatch was demonstrated with a commercial 1300 nm spectrometer-based FD-OCT (Telesto, Thorlabs GmbH, Germany) with an axial resolution of 10 µm in air, which was set to an acquisition speed of 5.5 kHz A-scan rate. GVD mismatch between reference and sample arm was created by a 16 mm rod of SF57 in the reference arm.

For experiments at 900 nm and *in vivo* measurements of the human retina, a commercial high-resolution spectrometer-based FD-OCT (Ganymede, Thorlabs GmbH, Germany) was used. It provides an axial resolution of 4 µm in air with an acquisition speed of 29 kHz. For measurements of the posterior segment of the human eye, it was coupled to a slit-lamp by using the scanning unit and reference of a commercial FD-OCT (SL SCAN-1, Topcon Europe, Netherlands).

4.5 Results and discussion

4.5.1 Axial motion in full-field SS-OCT

The CCSBR algorithm was able to correct for dynamically changing dispersion in full-field SS-OCT, which was introduced by sample motion during the wavelength sweep. Imaging of the finger tip with 36 volumes/second showed considerable broadening of the surface reflection and a wash-out of the sweat glands, even when the finger was stabilized in a ring (Figure 4.5.1a). For motion correction windows of 64 pixel at every 16th pixel were used. The apparent change of the axial position Δz over the sweep, which was determined by cross-correlation of the STFTs, was more than 14 pixel which corresponds to 100 µm (Figure 4.5.1c). This is about 17 times the real axial movement. After calculation of $\phi(k)$ and correction no motion artifacts were visible in the reconstructed image (Figure 4.5.1b). The CCSBR algorithm improved the FWHM from 4.3 pixel to 2.1 pixel, which is almost the theoretical possible optimum for a Hann windowed signal of 2.0 pixel for a flat reflecting surface [106]. However, the improvement of the image quality appears to be even larger. According to the extracted shift Δz (Figure 4.5.1c) only

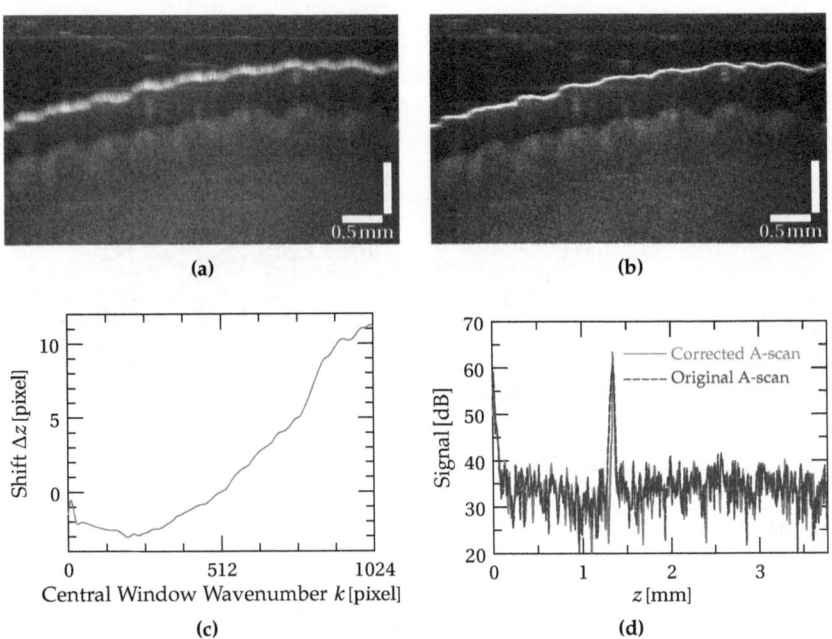

Figure 4.5.1.: B-scan from a volumetric full-field SS-OCT image of the finger tip. For one volume scan, images at 1024 wavelengths were acquired with 36,000 fps. Although the finger was stabilized by a ring, the acquisition time of 28 ms was not short enough to prevent motion induced blurring (a). Multiplication prior to the Fourier transform with the phase correction, which was determined by CCSBR, restores the original resolution and enhances the sharpness of the image (b). The apparent depth of the image structure represented as extracted motion curve shows changes of over 14 pixels during the scan (c). A-scans across the tissue surface in the original and the corrected images show the effect of motion compensation of the signal (d).

an approximately $5 - 10\times$ higher imaging speed could achieve an imaging without motion correction.

Still, motion correction using CCSBR cannot restore the resolution if the motion is too severe. When the imaging speed was reduced to nearly 4 volumes/second, the corrected images did not attain the full resolution and contained severe image artifacts (Figure 4.5.2). During the sweep the apparent position of structures in the image moved over 100 pixel which corresponded to 0.74 mm ($\approx 1000\lambda$) axial shift. Not only the amplitude of the motion was higher, also the movement was not monotonously and changed the direction five times over the wavelength sweep, which took about 260 ms. The STFT was no longer able to resolve the temporal

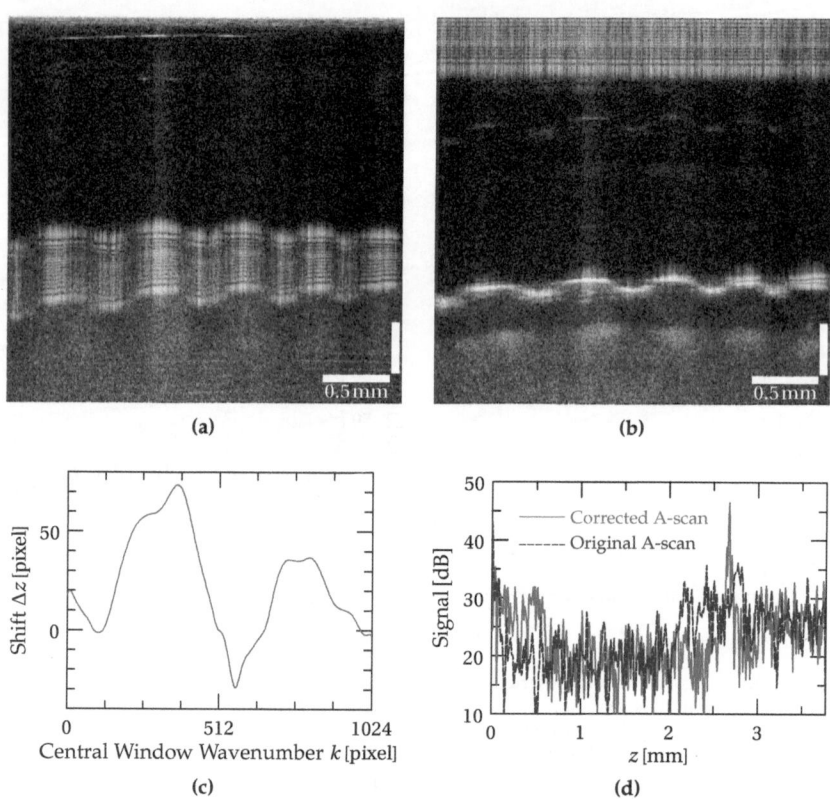

(a)   (b)

(c)   (d)

Figure 4.5.2.: B-scan from a volumetric full-field SS-OCT image of the finger tip which was measured at 4000 fps in 260 ms. The finger was also stabilized by a ring. Severe artifacts and blurring are visible (a). Though the numerical motion reduction improves image quality, it was not able to fully restore image quality (b). An apparent z-movement of nearly 100 pixels is observed (c). A-scans from the original and the corrected data show the loss of the depth resolution (d).

changes of the frequency chirp and CCSBR could not recover $\phi(k)$ with sufficient accuracy. Though not restoring the full image quality, the FWHM of the finger surface reflection was about 2.1 pixel after the correction, again close to the optimal FWHM of 2.0 pixel. Additionally, nonuniform axial blurring in the uncorrected and lateral blurring in the corrected image is observed. This indicates, that lateral motion and/or rotation might have contributed to the image degradation.

The presented examples demonstrate the dominance of axial motion on the image quality due to the magnification by the Doppler effect. As the axial and lateral resolution increase this will change, because $\lambda_f / \left(\lambda_i - \lambda_f \right)$ is reduced and lateral motion gets more visible (compare Section 4.1.1 and [98]). As soon as lateral motion can no longer be neglected the correction fails. The proposed algorithm to correct motion induced phase errors is computational expensive compared to full-field SS-OCT processing without compensation. But it allows to reduce acquisition speed and reduces setup costs. On a dual 2.13 GHz Quad Core Intel Xeon processor, the algorithm takes approximately 3 minutes for a complete measurement of $\phi(k)$ of a full-field raw volume of size $512 \times 192 \times 1024$. Performance could probably be improved by restricted the measurement to a local subregion of the volume or by porting it to a graphics processing unit (GPU) which has been shown to increase OCT processing speed [107, 108].

4.5.2 GVD mismatch in FD-OCT

CCSBR was also tested for correction of GVD mismatch. Windows of 128 pixel, centered at every 8th pixel were used for the STFT, which was applied to B-scan data with 2048 lateral positions. Each scan position had a spectrum with 1024 pixel for the 1300 nm FD-OCT and 2048 pixel for the 900 nm high-resolution FD-OCT, respectively. The data were fitted by a 4th degree polynomial to recover the phase function.

In vivo measurements of a finger tip were performed with a GVD mismatch of 1.9 fs/nm using the 1300 nm spectrometer-based OCT. When the conjugate of the recovered phase error was multiplied to the analytical interference signal prior to the FFT, the FWHM of the surface of the finger tip as shown in Figure 4.5.3 could be reduced from about 36 μm to about 13 μm. The improved FWHM corresponds to about 2.6 pixel, which is close to theoretical possible optimum of 2.0 pixel for a Hann windowed spectrum [106].

Additionally, *in vivo* measurements of the human retina were performed where the dispersion was only due to residual non-compensated dispersion of the human eye itself. Retina measurements were performed of a normal sighted and of a myopic eye (−15 diopter) which differ in the GVD due to the different length of the eye bulb. Although the GVD in the setup was matched for normal sighted eyes, imaging quality could still be improved even for the normal sighted eye as can be seen from Figure 4.5.4a and 4.5.4b, because higher orders of dispersion were not corrected by using glass substrates for the GVD matching. For the myopic eye the axial blurring of the point spread function due to dispersion is even better visible (Figure 4.5.4c). Correcting the GVD of the myopic eye using the phase

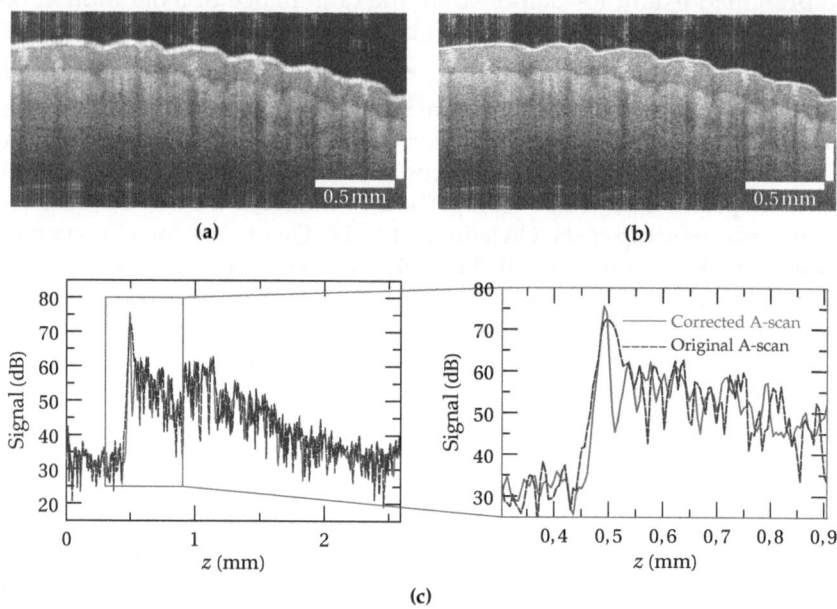

Figure 4.5.3.: OCT image of the finger tip measured with a scanning OCT with 16 mm of SF57 in the reference in order to create the GVD mismatch of 1.9 fs/mm. Considerable blurring of the surface reflection and the sweat ducts is observed (a). Using the correlation of sub-bandwidth reconstructed images to determine the correction function, image quality is completely restored (b). An improvement in the FWHM of the finger surface can be seen in the A-scan, especially when magnified between 0.3 mm and 0.9 mm (c).

error function $\phi(k)$ that was obtained for the normal sighted eye does not give optimal results as can be seen from Figure 4.5.4d. This shows for high-resolution OCT that the difference in GVD between the normal sighted and the myopic eye is sufficient to decrease imaging quality and an individual correction is required for full resolution. Using the individual phase error function $\phi(k)$ the B-scan provides the maximum improvement in axial resolution and imaging quality as can be seen from Figure 4.5.4e.

To compare the performance of CCSBR to a calibration measurement with a mirror according to Section 3.2.1 and [95], the improvement in the point spread function (PSF) was directly measured by covering a mirror with 20 mm of water. In both cases the PSF was improved significantly, showing hardly any difference between the two methods as shown in Figure 4.5.5a. Correspondingly, the phase

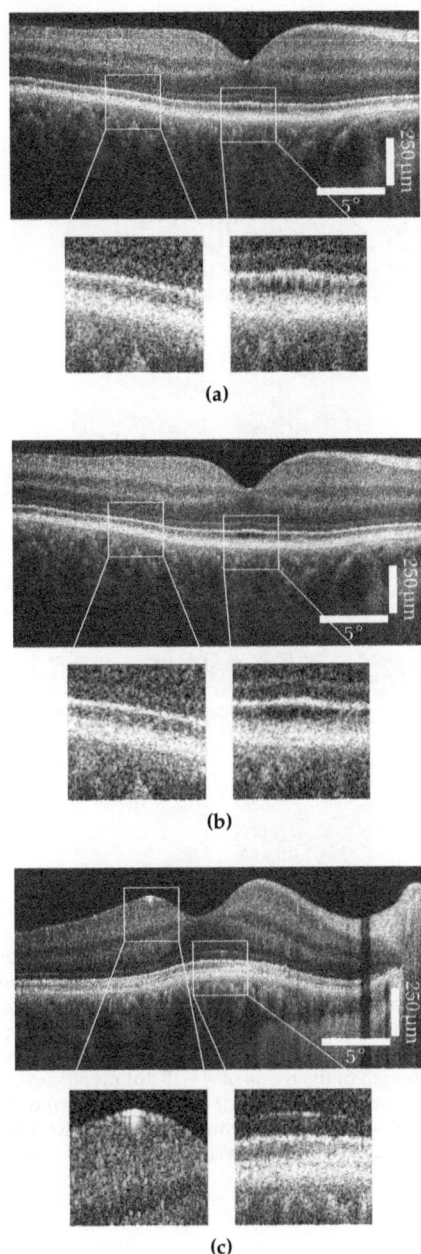

(a)

(b)

(c)

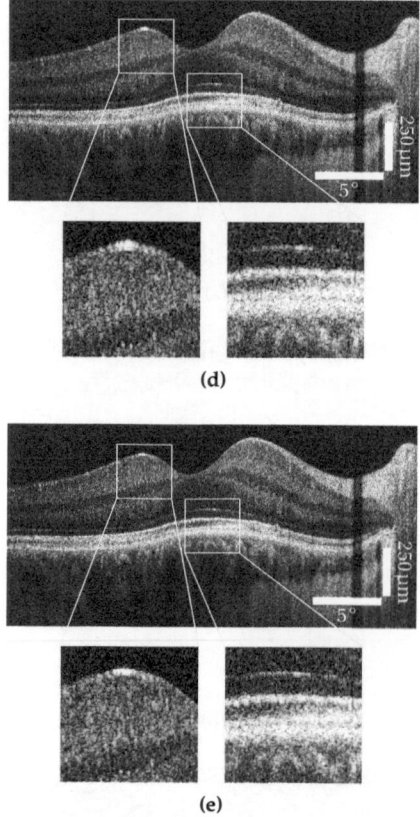

(d)

(e)

Figure 4.5.4.: Demonstration of the effects of GVD mismatch and its correction by CCSBR for *in vivo* ultra-high-resolution OCT of the retina. The GVD in the reference was matched for an average normal sighted person. a) Uncorrected B-scan of the retina of a normal sighted eye. Although the GVD mismatch is expected to be low, slight blurring of the axial structures is visible. b) Corrected image of the same data set. An improvement of the axial structures is visible. c) Uncorrected B-scan of the retina of a −15 dpt myopic eye. Image quality is severely degraded by blurring of axial structures. d) Corrected B-scan of the myopic eye using the phase errors $\phi(k)$ determined from the normal sighted eye. Residual blurring is still visible. e) Individual correction improves sharpness of the axial structures and borders.

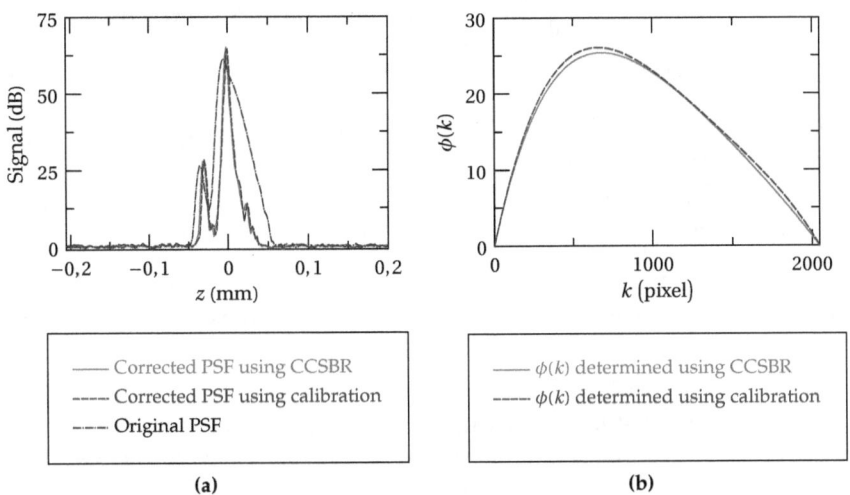

Figure 4.5.5.: Point spread function (PSF) obtained by imaging a mirror covered by 20 mm of water. a) A suitable calibration was obtained by imaging the mirror in two different positions [95] and the resulting GVD mismatch corrected PSF is compared to the PSF corrected by the phase function obtained using CCSBR and to the original uncorrected PSF. The PSFs obtained using calibration and CCSBR are basically indistinguishable. b) The obtained phase functions of both approaches.

errors $\phi(k)$ that were obtained using the two methods, show only minor difference (Figure 4.5.5b).

In general dispersion correction is quite robust using the CCSBR algorithm. The phase error $\phi(k)$ directly depends on the refractive index function $n(k)$, which varies according to the Kramers-Kronig relation in non-absorbing media only slowly with the wavenumber. Therefore in the optical window of tissue, $n(k)$ and $\phi(k)$ are continuous and smooth functions that can be well approximated by a low-degree polynomials.

The proposed algorithm shows good performance. A comparison with two alternatives, the calculation of phase error from previously measured PSFs and iterative improvement of image sharpness is shown in Table 4.5.1. The proposed CCSBR algorithm does not need a priory calibration measurements and is therefore especially suitable for correcting sample-induced GVD. Compared to approaches which parametrize the phase error function $\phi(k)$ and optimize the image quality, CCSBR is deterministic and should be faster. This makes the algorithm simpler and more effective. The described filtering for fixed pattern noise is not necessary for

	Any polyn. degree	No calibration	Non-iterative	No image evaluation
PSF phase retrieval [95]	Yes	No	Yes	Yes
Iterative improvement of image sharpness [104]	No	Yes	No	No
Cross-correlation of sub-bandwidth reconstructions	Yes	Yes	Yes	Yes

Table 4.5.1.: Properties of methods to determine GVD mismatch between sample and reference arm.

spectrometer-based FD-OCT. On a dual 2.13 GHz Quad Core Intel Xeon processor, the phase error detection algorithm took approximately $4-5$ s for FD-OCT spectral raw data of size 2048×1024 and approximately 10 seconds for raw data of size 2048×2048. Using GPUs [107, 108] the computation time for the phase error function might even be reduced to allow almost real-time determination. If the GVD is not dependent on the sample, the function $\phi(k)$ does not need to be determined for every measurement, as the obtained phase data will be valid for subsequent measurements.

5 Holoscopy

By parallel illumination and detection, swept-source full-field OCT enables high-speed tomographic imaging (Section 2.7.5 or [38,39]). Collecting all photons, that reach the area camera, it does no longer suffer from reduced imaging sensitivity, caused by the confocal gating. But still, the optimal lateral resolution is restricted by the depth of focus. Parallel acquisition of deep volumes is therefore not feasible for high-resolution imaging, as the depth of focus becomes significantly smaller than the imaging depth and FD-OCT looses its advantage compared to TD-OCT.

Digital holography (DH) is known for its capabilities to capture entire wave fields, including amplitude and phase (Section 2.6.2). It allows for numerical refocusing and does not suffer from a limited depth of focus. Therefore it promises to be the ideal approach to be combined with full-field FD-OCT to remove the restriction of FD-OCT to the Rayleigh range.

Here, it will be shown how to combine FD-OCT with DH to achieve a depth-independent lateral resolution. The resulting technique was named "holoscopy", a word created from the terms "holography" and "microscopy". This Chapter is based on the publications "Holoscopy — holographic optical coherence tomography" [109] and "Efficient holoscopy image reconstruction" [110].

5.1 Sensitivity improvement of holoscopy

Confocal imaging and scanning FD-OCT only work close to the focus of the detection beam, as the confocal gating rejects photons that are backscattered a few Rayleigh lengths z_R away. Detection of light that is reflected or scattered from a

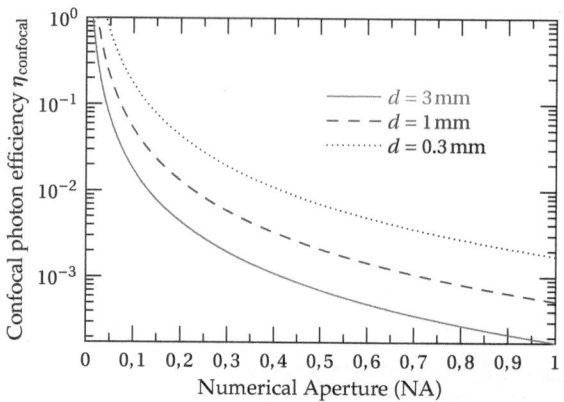

Figure 5.1.1.: Photon efficiency $\eta_{\text{confocal}} = 2z_R/d$ for a confocal imaging system such as scanning FD-OCT. It describes the relative amount of photons backscattered from the sample that can be used without refocusing for optimal imaging, i.e. that can be detected and deliver diffraction limited resolution.

depth of $\pm z_R$ from the focus layer will already be attenuated by 3 dB due to the confocal gating [111] and the lateral resolution is also degraded. The Rayleigh length is given by (2.4.14) to

$$z_R = \frac{\lambda}{\pi NA^2}.$$

The total measurement depth d will in general be larger and accordingly, as a rule of thumb, only a ratio of $2z_R/d$ of all photons, that are actually backscattered from the sample volume, can be used for imaging with optimal resolution and sensitivity. One can therefore define a photon efficiency of confocal imaging η_{confocal} by

$$\eta_{\text{confocal}} = \frac{2z_R}{d} = \frac{2\lambda}{\pi d NA^2}.$$

The photon efficiency for various measurement depths, ranging from 0.3 mm to 3 mm, and a central wavelength of 823.5 nm is shown in Figure 5.1.1 as a function of the NA. It can be seen that it drops rapidly with increasing NA, and for microscopic NA it is even several orders of magnitude smaller than the optimal value $\eta_{\text{confocal}} = 1$.

In holoscopy all photons backscattered from the sample can in principle be used optimally without refocusing and thus $\eta_{\text{holoscopy}}$ can reach a value of one, independent of measurement depth and imaging NA. It can be used to obtain optical tomographies more efficiently, and allows either to increase the sensitivity, to reduce the light intensity on the specimen and/or to increase the imaging speed by several orders of magnitude.

5.2 Basic setup

Holoscopy is based on interference. The sample is illuminated with an extended collimated beam and the backscattered or reflected light is superimposed with an extended, coherent and well-known reference beam. The resulting interference pattern is captured by a camera at many different wavenumbers. No imaging optics are required for this, but the achievable lateral resolution is limited without using imaging optics. Therefore two possible scenarios will be considered: lens-less holoscopy and high-resolution holoscopy. In the latter scenario, an objective is used to magnify the sample. A more detailed description of the setups that were actually used will be given in Section 5.6.3, schematic drawings are shown in Figure 5.2.1a and 5.2.1b.

5.3 Theory of holoscopy

The following Sections describe how the signals, as acquired by the camera, emerge in holoscopy, and how they can be efficiently processed to obtain tomographic images.

5.3.1 The intensity distribution on the camera

The image $I(x, y, t)$ on the area camera is given by a superposition of a reference wave field $R(x, y, t)$ and a sample wave field $O(x, y, t)$, where x and y give the respective lateral coordinate, i.e. the pixel of the camera and t is the time when the data was acquired. For theoretical considerations it is assumed that the fields R and O have been acquired for the entire plane, i.e. for all (x, y). In practice this is never the case, and limitations will result in a reduced aperture and a limited resolution (see also Section 5.4.3) and/or lateral field of view. Additionally, it is assumed that the camera lies in the $z = z_0$ plane. As in swept-source OCT, the wavenumber k at which a camera image is acquired will depend monotonically on the time t. The function $k(t)$ needs to be invertible on $t \in [0, T]$, i.e. a function $t(k)$ must exist, which always holds for a monotonic function with compact support. Ideally it will even be a linear function

$$k(t) = m_k t + k_i,$$

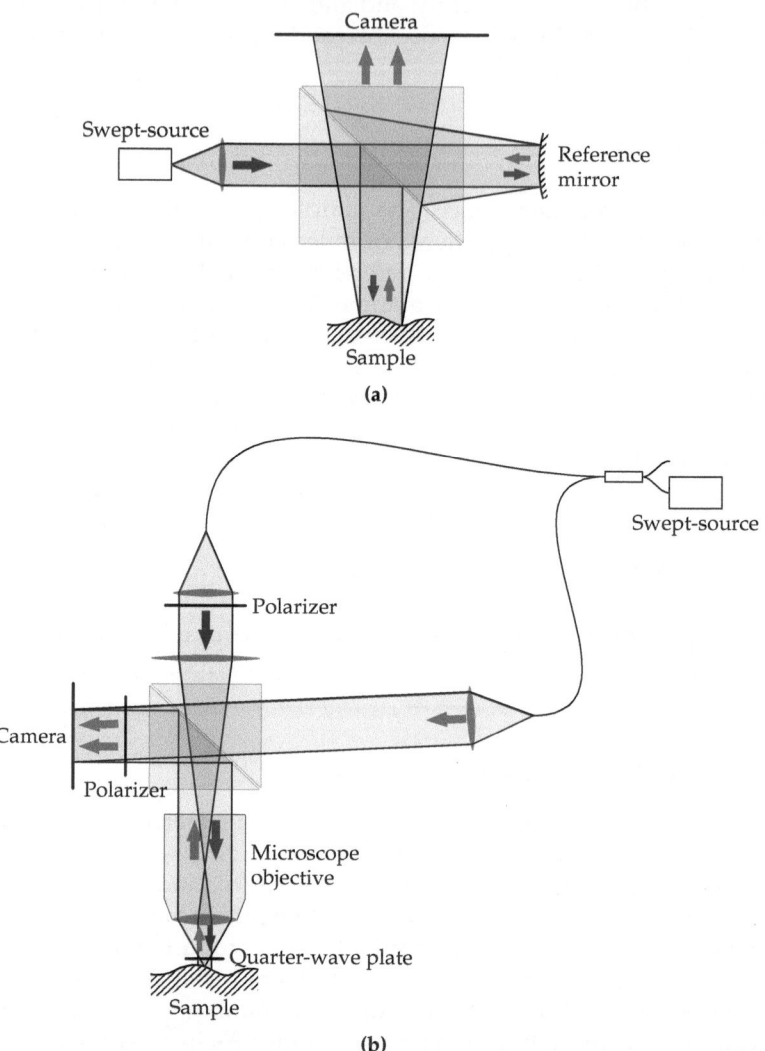

Figure 5.2.1.: a) Schematic drawing of a lens-less holoscopy setup. Coherent light of a known reference beam and light scattered from the sample are superimposed and digitized for many wavelengths. b) Mach-Zehnder type setup of a high-resolution holoscope. The sample light is brought onto the camera through a microscope objective, in order to resolve smaller structures. Polarization control and filtering is applied to suppress parasitic reflections.

with suitable sweep-rate m_k and initial wavenumber k_i. Without loss of generality, it is assumed, that a measurement starts at $t = 0$. In practice, the acquisition will be limited to a time range $[0, T]$ and the wavenumbers will be limited to a range $\left[k_i, k_i + m_k T = k_f\right]$ accordingly.

Because of (2.3.5), the signal measured by the camera $I(x, y, t)$ will be proportional to the squared superposition of the waves

$$
\begin{aligned}
I(x, y, t) &= \gamma |R(x, y, t) + O(x, y, t)|^2 \\
&= \gamma \Big(|R(x, y, t)|^2 + |O(x, y, t)|^2 + (R^*O)(x, y, t) + (RO^*)(x, y, t) \Big) \\
&= \gamma \Big(|R(x, y, t)|^2 + |O(x, y, t)|^2 + 2\mathrm{Re}(R^*O)(x, y, t) \Big), \quad\quad (5.3.1)
\end{aligned}
$$

where γ is a factor incorporating sensitivity of the camera and all scaling involved between squared field strength and measured values.

The signal (5.3.1) is very similar to the according signal in holography as shown in Section 2.6. The difference lies in the response of photographic material compared to the camera and in the time and therefore wavenumber dependence of the fields. The interpretation of the terms is identical: the term $|R(x, y, t)|^2$ describes the absolute value of the reference field, which contributes mostly to the DC part of the recorded interference pattern. The term $|O(x, y, t)|^2$ describes the interference of the object wave with itself. Finally, $2\mathrm{Re}(R^*O)(x, y, t)$ is the real cross-correlation term and contains the information of interest.

As shown in Section 2.3.2.1, wave fields can be expanded in a superposition of plane waves. The plane wave expansion/angular spectrum is given by their respective two-dimensional Fourier transforms, for the reference and sample wave fields this is

$$
\begin{aligned}
\tilde{R}(k_x, k_y, t) &= \mathscr{F}_{xy}[R(x, y, t)] \\
&= \int dx \int dy \, R(x, y, t) \cdot e^{-i(k_x x + k_y y)}
\end{aligned}
$$

and

$$
\begin{aligned}
\tilde{O}(k_x, k_y, t) &= \mathscr{F}_{xy}[O(x, y, t)] \\
&= \int dx \int dy \, O(x, y, t) \cdot e^{-i(k_x x + k_y y)}.
\end{aligned}
$$

The reference wave can have different shapes and origins, in most cases it will be a plane or a spherical wave. A plane reference wave can be described by

$$R(x,y,t(k)) = A_C(k(t))A_R \cdot \exp[i\mathbf{k}(t) \cdot \mathbf{x} + i\phi_0(t)]|_{\mathbf{x}=(x,y,z_0)}, \qquad (5.3.2)$$

where $\mathbf{k}$ is the wave vector and defines wavelength and propagation direction of the wave, $z = z_0$ is the camera plane, $A_C(k)$ is the relative amplitude spectrum and A_R describes the overall amplitude of the reference wave. $\phi_0(t)$ is the initial phase in the reference plane $z = 0$, in which the path length in the sample arm is the same as the one of the reference mirror. In this plane, both waves, reference and sample, have the same phase for all wavenumbers k. This plane is therefore also referred to as zero-delay plane. For on-axis holography and on-axis holoscopy, the reference wave is propagating perpendicular to the camera making the reference wave a complex constant value independent of x and y. To reduce spatial fringe frequencies on the camera, a spherical reference wave can be used. This will be described in more detail in Section 5.5.

In case of the Michelson-type setup, as shown in Figure 5.2.1a, the spherical wave can be created by subjecting a plane wave to a reference mirror with a given focal length f. In a Mach-Zehnder type setup, a spherical wave can be created by focusing the light with a suitable lens. The following will describe the Michelson-type setup, but adjusting the formalism for a Mach-Zehnder type setup is straight forward by renaming or substituting a few variables. Additionally, the introduction of a spherical reference wave in our computations is equivalent to introducing a numerical lens, since both are achieved by identical phase multiplications (compare Section 2.4.3, especially (2.4.7)). For this reason an interference pattern acquired of a far-field object wave with a collimated reference wave (e.g. Mach-Zehnder type setup, Figure 5.2.1b) is reconstructed in the same way as an interference pattern generated by the Michelson-type setup (Figure 5.2.1a). Only the interpretation of the spherical reference wave multiplication is a numerical lens in the former case and the application of the actual reference wave in the latter case (see also Section 5.5).

Let the distance from the reference mirror to the camera be denoted z_0. Then the reference field is given by

$$R(x,y,t(k)) = A_C(k)A_R \cdot e^{ik\sqrt{x^2+y^2+(z_0+f)^2}-ikf+i\phi_0(t)}. \qquad (5.3.3)$$

This describes a spherical wave originating at a distance f behind the reference plane (Figure 5.3.1).

For the sample, one assumes that the fraction of light backscattered at a transversal position (x,y) at a distance z from the reference plane is given by the scattering

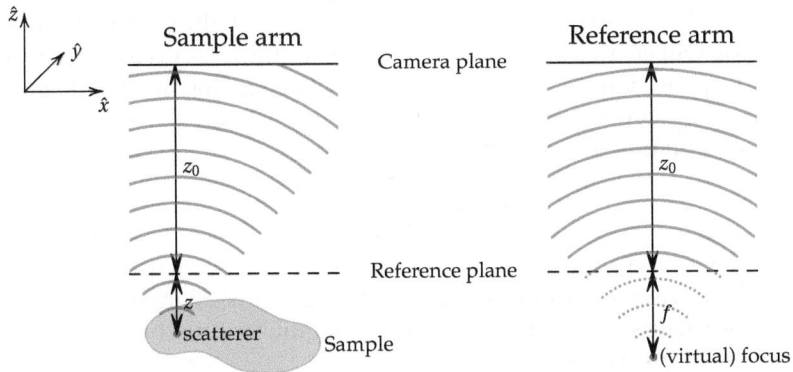

Figure 5.3.1.: Coordinate systems as used in the computation of the sample (left) and reference (right) wave field. The sample is made of several point scatterers whose fields are superimposed in the camera plane. In this case, the reference wave is a spherical wave with a (virtual) origin behind the reference plane. The configuration can be achieved by using a spherical reference mirror as shown in Figure 5.2.1a.

potential $\eta(x, y, z)$, thereby neglecting any angular dependence of the backscattering. The collimated light enters the sample, moves a distance z to the scatterer, is backscattered and then propagated by a distance $z + z_0$ to the camera. Assuming the validity of the first order Born approximation (see Section 2.5.2), completely neglecting any attenuation or reflection, and assuming a constant incident wave field throughout the volume, the object field and its angular spectrum in the camera plane are just a superposition of the fields generated by the backscattering in each depth, i.e.

$$O(x, y, t(k)) = A_O A_C(k(t)) \int dz \, \mathscr{P}_{k,z_0+z}\left[\eta(x, y, z) e^{ikz} e^{i\phi_0(t)}\right], \qquad (5.3.4)$$

$$\tilde{O}_{xy}(k_x, k_y, t(k)) = A_O A_C(k(t)) \int dz \, e^{ik_z(z+z_0)+ikz} \cdot e^{i\phi_0(t)} \cdot \tilde{\eta}_{xy}(k_x, k_y, z).$$

These assumptions are in generally also used in OCT (Section 2.7). The coordinate system as used for reference and wave field here is also illustrated in Figure 5.3.1.

5.3.2 The phase-corrected propagator, object and reference field

Propagating a wave field by the propagator $\mathscr{P}_{k,z}[\cdot]$ as shown in Section 2.3.2.2 will change the overall phase of the field. In OCT, depth information is contained in the phase of the wave field and thus changing the phase will change or destroy this depth-coding. If a wave is propagated by a distance z by the propagator $\mathscr{P}_{k,z}[\cdot]$, its overall phase will change by e^{ikz}. If the focusing of the object is to change, without changing the phase information, and thus without changing the depth encoding of the OCT part, the propagator needs to be adapted accordingly. This motivates the definition of a phase-corrected propagator as

$$\mathscr{P}^0_{k,z}[\cdot] \equiv e^{-ikz}\,\mathscr{P}_{k,z}[\cdot].$$

The phase-corrected propagator is just a mathematical construct, physical propagation of the wave field in free space will inevitably change the phase of the wave. Nevertheless, computations simplify significantly when introducing the phase-corrected propagator.

Considering the Fresnel approximation of the phase-corrected propagator, the additional phase term cancels the constant phase term of the original Fresnel kernel, resulting in a very simple representation of the corrected propagator kernel $P^0_{k,z}(k_x, k_y)$. It is given by

$$P^0_{k,z}(k_x, k_y) = e^{-i\frac{z}{2k}(k_x^2 + k_y^2)} = Q_{-z/k}(k_x, k_y). \tag{5.3.5}$$

The reference as well as the object wave field, given by (5.3.3) and (5.3.4), travel the same optical path length from the light source to the reference plane and have an identical time-dependent phase term $\phi_0(t)$, and common phase factors occur in reference and sample that need to be taken into account during reconstruction. In general, changing the overall phase of the two fields in exactly the same manner does not change the measurable quantity $I(x, y, t)$. For the following computations, it is therefore advantageous to redefine and simplify the phases of object and reference field, instead of using the previously obtained and physically motivated formulas, similar to the way the phase-corrected propagator replaces the propagator (2.3.6).

The phase-corrected reference wave field is therefore introduced by

$$\begin{aligned}
R_0(x, y, k) &\equiv R(x, y, t(k)) \cdot e^{-i\phi_0(t(k))} \cdot e^{-ikz_0} \\
&= A_R A_C(k) e^{ik\sqrt{x^2+y^2+(z_0+f)^2} - ik(z_0+f)}.
\end{aligned} \tag{5.3.6}$$

For $f \to 0$ the origin of the reference wave goes to the reference plane. Holograms of this kind are also referred to as Fourier-Holograms as they can be reconstructed in paraxial approximation by means of a simple Fourier transform (see e.g. [52]). In the paraxial approximation of the phase-corrected reference wave field R_0 phase-only terms cancel, resulting in a very simple expression. It is achieved by Taylor expansion of the argument of the $\exp(\cdot)$ function up to its quadratic terms:

$$R_0(x,y,k) \approx A_R A_C(k) e^{i\frac{k}{2(z_0+f)}(x^2+y^2)} = A_R A_C(k) Q_{k/(z_0+f)}(x,y). \qquad (5.3.7)$$

The phase-corrected object wave field is accordingly defined by

$$
\begin{aligned}
O_0(x,y,k) &\equiv O(x,y,t(k)) \cdot e^{-i\phi_0(t(k))} \cdot e^{-ikz_0} \\
&= A_O A_C(k) \int dz\, e^{-ikz_0} \mathscr{P}_{k,z_0} \mathscr{P}_{k,z} \Big[\eta(x,y,z)e^{ikz}\Big] \\
&= A_O A_C(k) \int dz\, \mathscr{P}^0_{k,z_0+z} \Big[\eta(x,y,z)e^{i2kz}\Big]. \qquad (5.3.8)
\end{aligned}
$$

It is worthwhile to note the similarity to the standard FD-OCT cross-correlation term of (2.7.2), except for the phase-corrected propagator $\mathscr{P}^0_{k,z_0+z}[\cdot]$. If the effect of the propagator can be partly or fully reverted, images can be reconstructed in analogy to the FD-OCT case. The effect of a propagator also arises in standard FD-OCT, but its influence is neglectable at relatively low NA and low imaging depth.

5.3.3 Obtaining the phase-corrected object wave field

The phase-corrected object and reference wave are equally well suited to describe the measured quantity. For the modified fields, the relation

$$I(x,y;k) = \gamma|R(x,y,k) + O(x,y,k)|^2 = \gamma|R_0(x,y,k) + O_0(x,y,k)|^2 \qquad (5.3.9)$$

holds and consequently also (5.3.1) is still true, if R and O are replaced by R_0 and O_0, respectively.

5.3.3.1 Reducing the DC signals

In order to perform a reconstruction of the backscattering potential $\eta(x,y,z)$ the phase-corrected object wave field $O_0(x,y,t)$ needs to be extracted from the measured intensity $I(x,y,t)$. To obtain it, the DC-parts need to be removed from the measured intensity. Assuming that the modulation frequencies are high and phases and lateral frequencies change from image to image, an averaged image $\bar{I}(x,y)$ as given by

$$\bar{I}(x,y) = \frac{1}{T} \int_T dt\, I(x,y,t)$$

can be removed from each acquired image as the modulation signals cancel in the averaging operation, i.e. high modulation frequencies will cause phase-washout and thus only the DC parts remain. The same can be done for an average spectrum $\bar{I}(t)$:

$$\bar{I}(t) \;=\; \frac{1}{A} \iint_A dx\, dy\, I(x,y,t) \propto S(k(t))$$

In order to avoid overcompensating the average intensity $\bar{I}$ given by

$$\bar{I} = \frac{1}{AT} \iint_A dx\, dy \int_T dt\, I(x,y,t)$$

needs to be added additionally. Both, the average image $\bar{I}(x,y)$ and the average spectrum $\bar{I}(t)$ can also be used to deconvolve the data and thus remove artifacts due to lateral or axial shaping of the acquired data. Additionally, the dependency of the signals on the spectrum $S(k)$ is removed this way.

After this deconvolution a three-dimensional window function $w(x,y,t)$ should be multiplied to suppress side lobes due to the rectangular window that is present otherwise (see also Appendix B.1 and B.2). The corrected intensity field is then given by

$$I_{\text{corr}}(x,y,t) = w(x,y,t) \cdot \frac{I(x,y,t) - \bar{I}(x,y) - \bar{I}(t) + \bar{I}}{\bar{I}(x,y) \cdot \bar{I}(t)}, \qquad (5.3.10)$$

albeit care needs to be taken to avoid dividing by zero.

5.3.3.2 On-axis geometry

Separating image and twin image   For the corrected intensity field $I_{\text{corr}}(x,y,t)$ it can now be assumed, that it does no longer contain the DC parts and it is therefore given by

$$
\begin{aligned}
I_{\text{corr}}(x,y,t) &\approx \gamma\Big[|O_0|^2(x,y,t) + 2\mathrm{Re}(R_0{}^*O_0)(x,y,t)\Big] \\
&= \gamma\Big(|O_0|^2 + R_0^*O_0 + (R_0^*O_0)^*\Big)(x,y,t).
\end{aligned}
$$

The intensity signal thus contains two parts $R_0^*O_0$ and $(R_0^*O_0)^*$, which are conjugate to each other and which are referred to as twin images in holography. In digital holography, off-axis reference illumination is commonly used to shift these images laterally and distinguish them. In FD-OCT the two conjugated terms cause the inability to distinguish negative from positive time delays (compare Section 2.7.2).

If all object path lengths are longer than the reference path length, the complex signal can be obtained by filtering negative frequency components (i.e. by performing a Hilbert transform $\mathcal{H}_t[\cdot]$ with respect to the time axis, see Section 2.1.1)

$$\begin{aligned} I_{\text{corr}}^{\mathcal{H}}(x,y,t) &= I_{\text{corr}}(x,y,t) + i\mathcal{H}_t[I_{\text{corr}}(x,y,t)]. && (5.3.11) \\ &\approx \gamma(R_0{}^*O_0)(x,y,t) \\ &\quad + \gamma\left(|O_0|^2(x,y,t) + i\mathcal{H}_t\left[|O_0|^2(x,y,t)\right]\right). \end{aligned}$$

The phase-corrected object field can be approximately obtained by simply dividing the intensity (5.3.17) by the complex conjugated phase-corrected reference field (5.3.6):

$$\gamma O_0(x,y,t) \approx \frac{I_{\text{corr}}^{\mathcal{H}}(x,y,t)}{R_0^*(x,y,k(t))} + \text{autocorrelation term.} \qquad (5.3.12)$$

In practice, the Hilbert transform in general does not need to be performed. A separation of positive and negative frequencies is in principle achieved automatically by an axial Fourier transform later (Section 5.4.1), that is required for depth discrimination.

5.3.3.3 Off-axis geometry

The off-axis geometry is frequently used in DH and digital holographic microscopy (DHM) to achieve separation of image and twin image [52,53], except for the use of multiple wavelengths.

Separating image and twin image In off-axis geometry, the reference wave is given by (5.3.2), with k not being perpendicular to the camera plane $z = z_0$. Introducing the directional vector of k as $\hat{k} = \left(\hat{k}_x, \hat{k}_y, \hat{k}_z\right)$, one can rewrite the reference wave in the camera plane as

$$\begin{aligned} R(x,y;k) &= \left. A_R A_C(k)e^{ik(\hat{k}\cdot x)+i\phi_0(k)}\right|_{z=z_0} \\ &= A_R A_C(k)e^{ik(\hat{k}_x x+\hat{k}_y y+\hat{k}_z z_0)+i\phi_0(k)}. \end{aligned}$$

To make the following paragraph easier readable, functional dependencies on the variables x, y, k, k_x, and k_y will be dropped, unless they are necessary for clarity. By introducing the perpendicular reference wave $R_\perp = A_R A_C \exp(ikz_0 + i\phi_0)$, which is perpendicular to the camera plane, the xy-phase factors S_{xy} and the z-phase factor S_z, one can rewrite the reference wave R as

$$R = R_\perp S_{xy} S_z, \tag{5.3.13}$$

with

$$S_{xy} = e^{ik(\hat{k}_x x + \hat{k}_y y)} \quad \text{and} \quad S_z = e^{ik(\hat{k}_z z_0 - z_0)}.$$

These two fields S_{xy} and S_z have the properties

$$S_{xy}^* S_{xy} = 1 \quad \text{and} \quad S_z^* S_z = 1.$$

By using the shift properties of the Fourier transform (A.3.3) it follows

$$\mathscr{F}\left[S_{xy}f(x,y)\right] = \tilde{f}\left(k_x - k\hat{k}_x, k_y - k\hat{k}_y\right),$$
$$\mathscr{F}\left[S_z f(z)\right] = e^{ik(\hat{k}_z z_0 - z_0)} \tilde{f}(k_z).$$

The intensity observed on the camera is given by (5.3.9):

$$I = \gamma\left(|R_0|^2 + |O_0|^2 + R_0^* O_0 + O_0^* R_0\right). \tag{5.3.14}$$

The phase-corrected reference field R_0 can be written in terms of a phase-corrected perpendicular reference wave $R_{\perp,0} = A_R A_C$ as

$$R_0 = R_{\perp,0} S_{xy} S_z.$$

Inserting this into (5.3.14), one obtains

$$I = \gamma\left(|R_{\perp,0}|^2 + |O_0|^2 + R_{\perp,0}^* S_z^* S_{xy}^* O_0 + R_{\perp,0} S_z S_{xy} O_0^*\right),$$

and its two-dimensional Fourier transform

$$\frac{1}{\gamma}\mathscr{F}[I] = |R_{\perp,0}|^2 \delta(k_x, k_y) + \mathscr{F}\left[|O_0|^2\right]$$
$$+ R_\perp^* S_z^* \tilde{O}_0\left(k_x + k\hat{k}_x, k_y + k\hat{k}_y\right) + R_\perp S_z \tilde{O}_0^*\left(k_x - k\hat{k}_x, k_y - k\hat{k}_y\right).$$

In frequency space, the four terms are separated (compare Figure 5.3.2), if the reference wave vector $\boldsymbol{k}$ is suitably chosen.

A spatial band-pass filter selects the object wave and separates it from the other

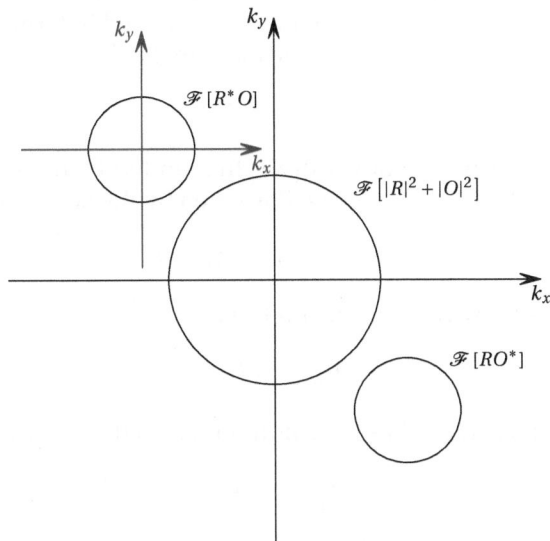

Figure 5.3.2.: Off-axis separation of the different interference terms in frequency space. Suitable filtering and an inverse Fourier transform can filter the appropriate wave field. The exact filter is wavelength dependent and thus a suitable phase shift needs to be added to compensate for this effect.

terms. A frequency space coordinate shift to center the object wave in frequency space removes all effects of the off-axis geometry on the object wave.

In practice, it is important to note, that $k_x = k\hat{k}_x$ and $k_y = k\hat{k}_y$ do change with the absolute wavenumber k, but the detection using the camera remains at fixed lateral frequencies $1/\Delta x$ and $1/\Delta y$ with Δx and Δy being the camera pixel spacing. The position of the object field in the Fourier transform of the image therefore depends on the wavenumber.

Spatial frequencies and sampling of cross and autocorrelation term   An object wave has a certain spacewidth, depending on the field of view it is representing, which will be denoted X. It also has a certain bandwidth, depending on its resolution. As shown in Section 2.4.4.1, the far field and the focused field of a wave field are related by a two-dimensional Fourier transform. Thus, the far-field can be identified with the angular spectrum of the focused field. The numerical aperture of the imaging optics limits the bandwidth of the angular-spectrum and thus of the far-field. The lateral bandwidth K of the object wave field, i.e. the maximal lateral frequencies, of the acquired image can thus be expressed in terms of the NA of the image and the wavenumber k outside the medium by

$$K = k \cdot \text{NA}.$$

Using a reference wave, with bandwidth K_R, the bandwidth of the cross-correlation term R^*O and also of its twin image O^*R is given by the sum of both

$$K_{R^*O} = K_{O^*R} = K_R + K.$$

For a parallel reference wave $K_R = 0$ and thus

$$K_{R^*O} = K.$$

The autocorrelation term $|O|^2$ has the bandlimit $2K$ as the bandwidth of O adds to itself

$$K_{|O|^2} = 2K.$$

For the plane reference wave it follows that

$$K_{|R|^2} = 0.$$

Band and spacelimit of the camera In one dimension, a camera with pixel-spacing Δx and N pixels has according to (2.2.2) a bandwidth

$$K_{\text{camera}} = \frac{N}{2} \cdot \Delta k = \frac{\pi}{\Delta x},$$

with Δk being the frequency between two adjacent pixels after the Fourier transform. Acquisition is in general also limited in its position space by the complete area of the camera, given by

$$X_{\text{camera}} = N\Delta x.$$

Bandlimits of focused images When imaging the object wave onto the camera with a magnification M, the effective camera spacewidth and bandwidth change according to

$$K_{\text{camera}} = M\frac{N}{2} \cdot \Delta k = M\frac{\pi}{\Delta x} \quad \text{and} \quad X_{\text{camera}} = N\frac{\Delta x}{M}.$$

To ensure correct imaging, the bandwidth of the camera K_{camera} must be larger than the bandwidth K of the image and the size of the camera spacewidth must be larger than the field of view

$$K_{camera} \gtrsim K \quad \text{and} \quad X_{camera} \gtrsim X.$$

In off-axis scenarios, one also needs to separate the autocorrelation term (radius $2K$) from the image term (radius K), which gives for a 45° tilt of the object-reference beam plane with respect to the camera axes, the condition

$$K_{camera} \gtrsim \left(\frac{3}{2}\sqrt{2}+1\right)K,$$

as demonstrated in Figure 5.3.3. In this case, it is the camera that limits the image field of view and the pixel spacing, i.e. the bandwidth of the camera, that limits resolution if chosen inappropriately. In general smaller off-axis angles or larger bandwidth (i.e. resolution) of the object wave can be used, but the autocorrelation and cross-correlation terms will overlap, resulting in coherence noise in the holoscopic data.

Bandlimits of far-field images When acquiring data in the far-field, as in the high-resolution setup shown in Figure 5.2.1b, the roles interchange. The camera size can limit the sampling of the far-field and therefore of the angular spectrum. The frequency space limits the acquisition of the object field of view. According to Section 2.4.4.2, the roles of X and K interchange according to

$$fK \leftrightarrow kX,$$

with f being the focal length of the objective and k being the current wavenumber. The bandwidth of the wave field in the camera plane is now given by the object field spacelimit to

$$K_{camera} \gtrsim \left(\frac{3}{2}\sqrt{2}+1\right)\frac{k}{f}X.$$

For the camera restriction it follows that

$$X_{camera} \gtrsim \frac{f}{k}K = f \cdot NA,$$

which basically states that the overall size of camera must not be the limiting aperture. However, this restriction is only valid directly behind the objective. Farther away, beams will diverge and might still miss the camera.

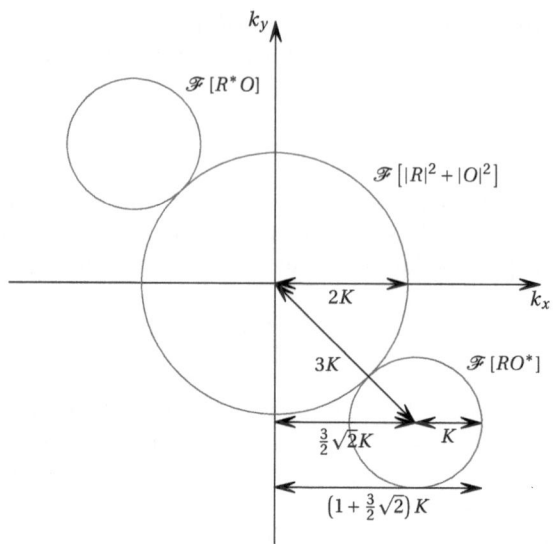

Figure 5.3.3.: Bandwidth of the interference terms for off-axis holographic or holoscopic data. The autocorrelation term has twice the bandwidth of the image terms. In order for the camera to resolve the pattern, the bandwidth of the camera needs to be larger than $\left(1 + 3/2\sqrt{2}\right)K$, with K being the bandwidth of the object wave field.

Frequency space coordinate shift For further easy computation, one shifts the coordinate system of frequency space such that the object wave is centered in the origin. In order to achieve this one needs to shift the spectral coordinate system (k_x, k_y) according to

$$
\begin{aligned}
k_x &\rightarrow k_x \pm k\hat{k}_x \\
k_y &\rightarrow k_y \pm k\hat{k}_y
\end{aligned}
$$

with suitable signs. This coordinate transformation is dependent on the wavenumber k. This poses no problem in digital holography, but for holoscopy a suitable coordinate transformation needs to be accomplished for various wavenumbers. But in order to keep the coordinate transformation on pixel boundaries, one needs to introduce a k-independent transformation, for example

$$
\begin{aligned}
k_x &\rightarrow k_x \pm k_0\hat{k}_x \\
k_y &\rightarrow k_y \pm k_0\hat{k}_y.
\end{aligned}
$$

After this transformation the frequencies will still shift by $(\mp k_0 - k)\hat{k}_x$ for different wavenumbers k. In order to correct this shift, a phase multiplication in position space with a phase factor Φ

$$\Phi(x, y; k) = e^{-i(\mp k_0 - k)(x\hat{k}_x + y\hat{k}_y)} \tag{5.3.15}$$

can be used, which can be seen as a multiplication with a remaining off-axis reference wave. According to the Fourier shift theorem (A.3.3), this corrects the missing alignment of images in frequency space (Figure 5.3.4).

Frequency space filtering Frequency domain filtering is applied to separate autocorrelated parts, DC parts and twin image from the image and reduce overall noise. It is applied by

$$I_{\text{filtered}}(x, y; k) = \mathscr{F}_{xy}^{-1}\left[w(k_x, k_y)\mathscr{F}_{xy}[I(x, y; k)]\right] \tag{5.3.16}$$

with a suitable filter function $w(k_x, k_y)$. The filter function $w(k_x, k_y)$ will act as computational aperture and the lateral PSF in position space is determined by the filter function (see Section 5.4.3), a simple rectangular filter will lead to lateral side lobes of high-intensity point scatterers.

The resulting signal of the spatial filter can be approximated by

$$I_{\text{corr}}^{\text{filtered}} \approx \gamma(R^*O)(x, y, t).$$

Analogous to (5.3.11) the corrected intensity $I_{\text{corr}}^{\text{filtered}}(x, y, t)$ can also be described by the phase-corrected reference and object-field as given by (5.3.6) and (5.3.8), respectively. Neglecting the autocorrelation term, it is given by:

$$I_{\text{corr}}^{\text{filtered}}(x, y, t) \approx \gamma(R^*O)(x, y, t) = \gamma(R_0^*O_0)(x, y, k(t)) \tag{5.3.17}$$

As in (5.3.12), the object wave field can now be computed

$$\gamma O_0(x, y, t) \approx \frac{I_{\text{corr}}^{\text{filtered}}(x, y, t)}{R_0^*(x, y, k(t))}. \tag{5.3.18}$$

The off-axis setup geometry allows to select one of the two twin images without overlap. Images of objects at positive and negative distances from the reference plane are completely separated after filtering, i.e. the off-axis recording provides automatically full-range OCT and the symmetry of FD-OCT with respect to the zero-delay plane can be removed (compare Section 2.7.2). Additionally, the auto-

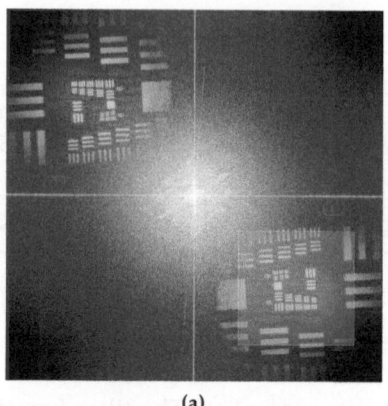

(a)

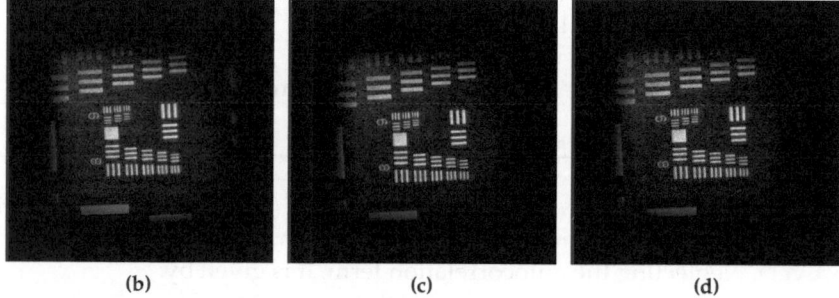

(b)   (c)   (d)

Figure 5.3.4.: a) Fourier transform of a far-field hologram of a US Air-Force (USAF) test chart acquired at 867.5 nm and the respective filtering window is shown. b) Reconstruction of the image of the hologram (a). c) Reconstruction of a hologram of the same object, acquired with the same setup, but at a wavelength of 816 nm. Compared to (b), the hologram does not only show a different field of view, but the complete image is shifted by a few pixels. d) Reconstruction of the same hologram as in (c) if the phase correction (5.3.15) is applied prior to the reconstruction. The hologram does still show a slightly different field of view compared to (b), but the common parts in both images overlay well. The dashed line is supposed to make the image shift more clearly.

correlation object term $|O_0(x, y, t)|^2$ is also separated, which is not possible with the on-axis geometry. The autocorrelation terms are in general not in focus if the object is focused, but they will add additional coherence noise, especially in the upper areas of the reconstructed volume.

5.4 Reconstruction

5.4.1 Reconstruction of one Rayleigh length

The phase-corrected object wave fields (5.3.12) or (5.3.18) for all wavenumbers – and thus all acquisition times – can be propagated back to a common plane of the volume. A phase-corrected propagation reverts the effect of the phase-corrected propagator in (5.3.8) for the chosen reconstruction distance. Depending on the chosen common plane of the volume, it additionally sets the focus of the reconstruction. The data obtained after this numerical computation, are equivalent to measured full-field FD-OCT data. A Fourier transform along the k-axis is performed, analogous to FD-OCT processing, to achieve the depth sectioning. The reconstructed volume will have diffraction limited resolution in the plane, where the acquired images have been propagated to. The other planes will have degraded lateral resolution as is observed in classical optics.

If one chooses a plane defined by the distance z_P relative to the reference plane, which is supposed to be reconstructed sharply, the reconstructed object is obtained by

$$
\begin{aligned}
\mathscr{P}^0_{k,-z_0-z_P}[O_0(x,y,k)] &= A_O A_C(k) \int dz \, \mathscr{P}^0_{k,-z_0-z_P} \mathscr{P}^0_{k,z_0+z} \left[\eta(x,y,z) e^{2ikz} \right] \\
&= A_O A_C(k) \int dz \, \mathscr{P}^0_{k,z-z_P} \left[\eta(x,y,z) e^{2ikz} \right],
\end{aligned}
$$

where (2.3.8) has been used to combine the operators. These data are now comparable to full-field FD-OCT data with the focus according to the chosen reconstruction depth $z = z_P$. Performing a Fourier transform with respect to the k-axis now gives the scattering potential

$$
\tilde{A}_C(z) * \eta_{z_P}(x,y,z) = \frac{1}{\pi A_O} \int dk \, \exp(-i2kz) \mathscr{P}^0_{k,-z_0-z_P}[O_0(x,y;k)], \quad (5.4.1)
$$

where $*$ denotes the convolution operation with respect to the z-axis and $\tilde{A}_C(z)$ is the point spread function obtained by Fourier transforming the amplitude spectrum $A_C(k)$. Due to (2.3.9) the scattering potential is only reconstructed without error for the layer $z = z_P$.

5.4.2 One-step reconstruction

5.4.2.1 In free space

In order to get a complete depth-independent reconstruction of the volume data, one needs to repeatedly perform a reconstruction by (5.4.1) for different reconstruction depths z_P: the procedure outlined in Section 5.4.1 needs to be repeated so that the propagator vanishes for all layers. In order to achieve this, one needs to set the chosen focus layer z_P to the actual depth z, i.e. $z_P = z$, in (5.4.1) which gives

$$\tilde{A}_C(z) * \eta(x,y,z) = \frac{1}{\pi A_O} \int dk\, e^{-i2kz} \mathscr{P}^0_{k,-z_0-z_P} [O_0(x,y,k)] \Big|_{z_P=z}.$$

In the angular spectrum, the respective equation reads

$$\tilde{A}_C(z) * \tilde{\eta}_{xy}(k_x,k_y,z) = \frac{1}{\pi A_O} \int dk\, e^{-i2kz} e^{-i(k_z-k)(z_0+z_P)} [\tilde{O}_0(k_x,k_y,k)] \Big|_{z_P=z}.$$

The convolution operation $*$ only affects the z-axis and thus remains untouched from the change to the angular spectrum. Finally, setting effectively $z_P = z$ gives

$$= \frac{1}{\pi A_O} \int dk\, e^{-i(k+k_z)z} e^{-i(k_z-k)z_0} [\tilde{O}_0(k_x,k_y,k)].$$

Going back to spatial space gives

$$\tilde{A}_C(z) * \eta(x,y,z)$$
$$= \frac{1}{\pi A_O} \mathscr{F}^{-1}_{xy} \left[\int dk\, e^{-i(k+k_z)z} e^{-i(k_z-k)z_0} \mathscr{F}_{xy}[O_0(x,y,k)] \right]. \quad (5.4.2)$$

With this reconstruction all layers are reconstructed with diffraction limited resolution. However, this relation is only valid if the free-space propagator applied here is correct. For example, lenses in between sample and camera or a sample with refractive index need a different propagator.

5.4.2.2 In a medium

In a medium with index of refraction n, the wavenumber k will be increased by a factor n. Assuming that entry and exit to the medium are parallel to the camera plane, the lateral components of $\mathbf{k}$ will remain unchanged, as they must not change abruptly when entering or leaving the medium. Therefore, for propagation in the medium, the wave vector needs to be changed according to

$$\mathbf{k}' = \begin{pmatrix} k_x \\ k_y \\ k_z' \end{pmatrix} = \begin{pmatrix} k_x \\ k_y \\ \sqrt{n^2 k^2 - k_x^2 - k_y^2} \end{pmatrix},$$

and it follows

$$\|\mathbf{k}'\| = nk,$$

where $\mathbf{k}'$ denotes the wave vector in the medium. In the reconstruction by (5.4.2) the kernel needs to be modified by replacing k and k_z by nk and k_z', respectively. However, assuming that z_0 describes the propagation distance to focus the reference plane, which is in a medium might not be equal to the physical distance, the phase factor does not need to be modified. The reconstruction is then given by

$$
\begin{aligned}
&\tilde{A}_C(z) * \tilde{\eta}(k_x, k_y; z) \\
&= \frac{1}{\pi A_O} \int d(nk) \underbrace{\exp\left(-i(k_z' + nk)z\right)}_{\text{kernel}} \underbrace{\exp\left(-i(k_z - k)z_0\right)}_{\text{phase}} \tilde{O}_0(k_x, k_y; k). \quad (5.4.3)
\end{aligned}
$$

This integral transform with the modified kernel can also be calculated by an NFFT. By using $\sqrt{1 - x^2} \approx 1 - x^2/2$ a simplification is possible for not too high NA: in paraxial approximation the term for k_z can be expanded and rewritten to

$$\exp\left(+i(k_z' + nk)z\right) \approx \exp\left(+i\left(2nk + \frac{1}{n}(k_z - k)\right)z\right).$$

By introducing the optical path length $z' = nz$ the kernel is modified to

$$= \exp\left((+i(2 - \zeta)k + \zeta k_z)z'\right), \quad (5.4.4)$$

with

$$\zeta = \frac{1}{n^2}. \quad (5.4.5)$$

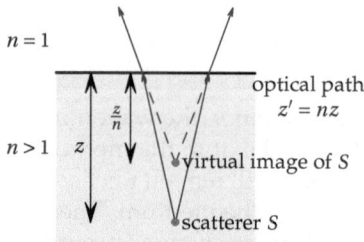

$n = 1$

$n > 1$ z

$\frac{z}{n}$

optical path
$z' = nz$

virtual image of S

scatterer S

Figure 5.4.1.: Due to refraction, the focus position is shifted approximately by a factor n in a medium compared to the according free-space focus.

ζ describes the proportionality constant between imaging distance and optical path length as illustrated in Figure 5.4.1. The complete reconstruction formula is

$$\tilde{A}_C(z') * \tilde{\eta}(k_x, k_y; z')$$

$$= \frac{1}{\pi A_O} \int dk \underbrace{e^{-i((2-\zeta)k+\zeta k_z)z'}}_{\text{kernel}} \underbrace{e^{-i(k_z-k)z_0}}_{\text{phase}} \tilde{O}_0(k_x, k_y; k). \quad (5.4.6)$$

For $\zeta = 0$, (5.4.6) reduces to (5.4.1) with $z_P = 0$, i.e. the chosen reconstruction distance is the reference plane. For increasing n the parameter ζ tends to zero, thus the higher the refractive index of the sample, the better the approximation by (5.4.1). In fact, for $\zeta < 0.5$ the reconstruction by (5.4.1) yields better results than the reconstruction by (5.4.2), which assumes free space. In case the paraxial approximation of (5.4.4) is not valid, the more specific formula by (5.4.3) needs to be used.

If n of the medium is not exactly known, an approximate solution of the reconstruction can be used for a fast experimental determination of ζ. By first Fourier transforming the object waves $O_0(x, y; k)$ with respect to the k-axis, i.e. using FD-OCT depth discrimination on the unprocessed holograms and only afterwards performing holographic refocusing with the center wavenumber for at least two different depths, the focus positions and the optical path lengths of these layers can be determined. A linear regression of these points gives ζ and the reference propagation length z_0.

5.4.2.3 With numerical magnification

There are several ways to magnify the lateral reconstruction of digital holograms, i.e. to change the spacing of lateral pixels (see e.g. [49, 52]). The easiest way to do this in lens-less holoscopy is to introduce either a magnification of the unprocessed holograms by a factor m or to change the origin of the reference wave from its original z-position z_{ref} to $z_{ref,rec}$. Another way of magnification is to change the reconstruction wavelength, which is mostly used in analog holography if the reconstruction wavelength is invisible, and is not considered here.

Assuming that the reference position that is used by the reconstruction is changed, but that the wavelengths for acquisition and reconstruction are identical, the transversal M_t and axial magnification M_a are given by

$$M_t = m \left| 1 - \frac{z_{obj}}{z_{ref}} \mp m^2 \frac{z_{obj}}{z_{ref,rec}} \right|^{-1} \quad \text{and} \quad M_a = M_t^2, \tag{5.4.7}$$

where m is the magnification of the not reconstructed hologram, z_{obj} is the distance of object to camera and z_{ref} is the distance of reference origin to camera. $z_{ref,rec}$ is the distance of the numerical reconstruction wave origin (see e.g. [49,50,52]). In order for this approximation to be valid it needs to be assumed, that z_{obj} is approximately constant over the complete measurement depth, i.e. the measurement depth is significantly smaller than the distant of the object plane to the 'virtual' camera plane. While this is the case for most lens-less setups (low NA), it is in general no longer true, if a high NA microscope objective is used to magnify the object wave field. For high NA holoscopy the magnification is thus not possible, and $z_{ref,rec}$ needs to be chosen properly as will be shown in Section 5.5.2. In the latter case a magnification can still be achieved by zero-padding in the frequency domain, which is most easily performed during spatial filtering by (5.3.16).

According to (5.4.7) the reference wave field needs to be adjusted in order to obtain a magnification. In the phase-corrected reference wave field (5.3.6) the reference origin is given by the reference distance and the focal length, i.e. $z_{ref} = z_{ref,rec} = z_0 + f$. By replacing R_0 by $R_{0,magn}$ that might be given by

$$R_{0,magn} = A_R A_C(k) e^{-i \frac{k}{2z_{rec,ref}} (x^2+y^2)} \quad \text{with } z_{ref,rec} = z_{ref}$$

a magnification can be introduced. The according object-wave field is then given by

$$O_{0,magn}(x,y,t) \approx \frac{I_{corr}^{\mathcal{H}/\text{filtered}}(x,y,t)}{\gamma R_{0,magn}^*(x,y,t)}.$$

For the single reconstruction the focusing step needs to be adjusted. However, the depth-reconstruction remains unchanged. This gives:

$$\tilde{A}_C(z) * \eta \left(\frac{x}{M_t}, \frac{y}{M_t}, z \right) = \frac{1}{\pi A_O} \int dk\, e^{+i2kz} \mathscr{P}^0_{k, -M_a z_0 - M_a z} \left[O_{0,\text{magn}}(x, y, t(k)) \right].$$

The full-reconstruction can then be computed accordingly to

$$\tilde{A}_C(z) * \eta \left(\frac{x}{M_t}, \frac{y}{M_t}, z \right) = \frac{1}{\pi A_O}$$
$$\times \mathscr{F}_{xy}^{-1} \left[\int dk\, e^{-i((2-M_a)k + M_a k_z)z} e^{-i(k_z - k) M_a z_0} \mathscr{F}_{xy} \left[O_{0,\text{magn}}(x, y, t(k)) \right] \right].$$

The interpolation function κ_ζ is thus now given by

$$\kappa_\zeta(k_x, k_y; k) = (2 - M_a)k + M_a k_z = (2 - \zeta)k + \zeta k_z$$

and therefore

$$\zeta = M_a. \tag{5.4.8}$$

By combining (5.4.5) and (5.4.8) one obtains a parameter ζ that incorporates the numerical magnification as well as the refractive index:

$$\zeta = \frac{M_a}{n^2} \tag{5.4.9}$$

5.4.2.4 The complete reconstruction integral

For each data point in the Fourier space at coordinates (k_x, k_y, k), one can compute the according k_z by

$$k_z(k_x, k_y; k) = \sqrt{k^2 - k_x^2 - k_y^2}.$$

One can define a function κ_ζ, which computes the actual interpolation variable:

$$\kappa_\zeta(k_x, k_y; k) = (2 - \zeta)k + \zeta k_z = (2 - \zeta)k + \zeta \sqrt{k^2 - k_x^2 - k_y^2} \tag{5.4.10}$$

Of special importance is also its inverse function $k(k_x, k_y; \kappa_\zeta)$, which is required for certain interpolation algorithms. It can be computed, for example by completing

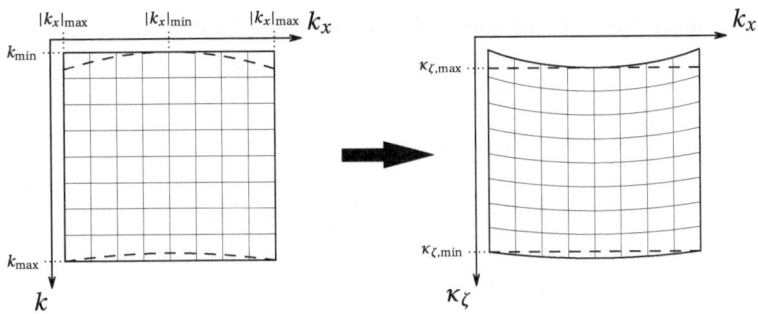

Figure 5.4.2.: Coordinate transformation in holoscopy, that is required for a one-step reconstruction to obtain the entire volume with diffraction limited resolution.

the square, to

$$k(k_x, k_y; \kappa_\zeta)$$
$$= \begin{cases} \frac{1}{2\kappa_\zeta}\left(\kappa_\zeta^2 + k_x^2 + k_y^2\right) & \text{for } \zeta = 1 \\ \frac{1}{4-4\zeta}\left[(2-\zeta)\kappa_\zeta + \sqrt{(2-\zeta)^2\kappa_\zeta^2 - (4-4\zeta)(\kappa^2 + \zeta^2 k_x^2 + \zeta^2 k_y^2)}\right] & \text{for } \zeta > 1 \quad (5.4.11) \\ \frac{1}{4-4\zeta}\left[(2-\zeta)\kappa_\zeta - \sqrt{(2-\zeta)^2\kappa_\zeta^2 - (4-4\zeta)(\kappa^2 + \zeta^2 k_x^2 + \zeta^2 k_y^2)}\right] & \text{for } \zeta < 1. \end{cases}$$

In practice, the integration over k is performed over a wavenumber subregion $K = \left[k_i; k_f\right]$ that is defined by the spectrum $S(k) = |A_C|^2(k)$, with k_i being the initial and k_f the final wavenumber during the sweep. Accordingly, there is a minimal and a maximal wavenumber $k_{\min}$ and $k_{\max}$, respectively. Additionally, there are maximal absolute spatial frequencies, depending on the camera, i.e. the size of the pixels corresponding to the Nyquist frequency (2.2.2). Let the set of spatial frequencies be given by K_X and K_Y. In general $(k_x)^2_{\min} = (k_y)^2_{\min} = 0$, assuming that $0 \in K_X$ and $0 \in K_Y$. The possible values for κ_ζ, that can be computed for a specific set of wavenumbers K and for all spatial frequencies $k_x \in K_X$ and $k_y \in K_Y$ is thus limited (see also Figure 5.4.2). It will be bounded by $\kappa_{\zeta,\max}$ and $\kappa_{\zeta,\min}$ which are given by

$$\kappa_{\zeta,\min} = (2-\zeta)k_{\min} + \zeta\sqrt{k_{\min}^2 - (k_x)^2_{\min} - (k_y)^2_{\min}} = 2k_{\min} \quad (5.4.12)$$

$$\kappa_{\zeta,\max} = (2-\zeta)k_{\max} + \zeta\sqrt{k_{\max}^2 - (k_x)^2_{\max} - (k_y)^2_{\max}} \quad (5.4.13)$$

The area of integration of suitable κ_ζ is thus restricted to

$$K_\kappa = \left[\kappa_{\zeta,\min}; \kappa_{\zeta,\max}\right].$$

The actual integral to be evaluated is

$$\int_K dk\, e^{-i\kappa_\zeta(k_x,k_y;k)z} f(k_x, k_y; k), \qquad (5.4.14)$$

where the function $f(k_x, k_y; k)$ is given by

$$f(k_x, k_y; k) = e^{-i(k_z - k)z_0} \mathscr{F}_{xy}[O_0(x, y, k)]. \qquad (5.4.15)$$

By a simple variable substitution the integration can be performed over κ_ζ instead of k and thus the integral can be efficiently evaluated by a fast Fourier transform. The required mapping is given by

$$(k_x, k_y, \kappa_\zeta)(k_x, k_y, k) = (k_x, k_y, ((2 - \zeta)k + \zeta k_z)).$$

The according integral over κ_ζ is then

$$\int_{K_\kappa} d\kappa_\zeta\, e^{-i\kappa_\zeta z} \left(\frac{dk}{d\kappa_\zeta}\right) f(k_x, k_y; k(\kappa_\zeta)).$$

Alternatively, the integral (5.4.14) can be computed using a non-equispaced Fourier transform (see Section 2.1.3, Chapter 3, and Appendix C.4.2).

Paraxial approximation　　In paraxial approximation (5.4.10) and (5.4.11) can be simplified significantly and in most cases they yield sufficient results. They can be computed to

$$\kappa_\zeta(k_x, k_y; k) = 2k - \zeta\left(\frac{k_x^2}{2k} + \frac{k_y^2}{2k}\right)$$

and

$$k = \frac{1}{4}\kappa_\zeta + \sqrt{\frac{\kappa_\zeta^2}{16} + \frac{\zeta}{4}\left(k_x^2 + k_y^2\right)},$$

respectively.

5.4.3 Resolution

5.4.3.1 Axial resolution

The axial resolution is determined by the sweep range of the swept-source laser and the spectral shape, after spectral apodization of the acquired signals. The resolution will thus be approximately identical to the OCT resolution using an identical light source (see e.g. [54]). However, applying the complete reconstruction, the interpolation variable κ_ζ is determining the axial resolution instead of the original wavenumber k. As some parts of the k-space cannot be used for reconstruction, it will thus be slightly decreased.

The discretization (see Section C.4.2) of the interpolated reconstruction integral shows that after interpolation, images will have a spacing of

$$\Delta\kappa_\zeta = \frac{\kappa_{\zeta,\max} - \kappa_{\zeta,\min}}{N},$$

with N being the number of images that were acquired during the sweep. After Fourier transform, the resulting spacing of subsequent en-face images will be

$$\Delta z = \frac{2\pi}{N\Delta\kappa_\zeta}.$$

According to Appendix B.1.2, the axial resolution (FWHM) z_{FWHM} for a Hann windowed spectrum will be twice as large, resulting in

$$z_{\text{FWHM}} = \frac{4\pi}{\kappa_{\zeta,\max} - \kappa_{\zeta,\min}}. \tag{5.4.16}$$

5.4.3.2 Lateral resolution

The lateral resolution is determined by the numerical aperture of the setup in analogy to classical imaging systems. The same formalism as for the axial resolution can be applied if the acquired holograms are apodized with a window function $w(x,y,z)$ in (5.3.10). In lens-less holoscopy the aperture is given by the shape of the camera area, assuming that all interference fringes are sampled correctly (Section 5.5.1). The resulting point spread function is given by (B.2.2).

For the setup using imaging optics (Figure 5.2.1b) the resolution is determined by the circular aperture of the objective resulting and an Airy point spread function is expected. The according lateral resolution is given by (B.2.1).

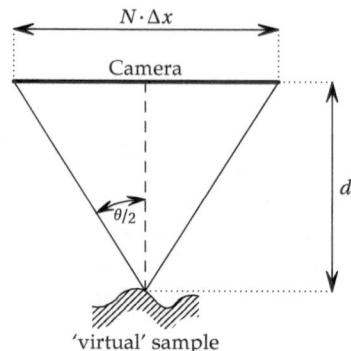

Figure 5.4.3.: Numerical aperture of a lens-less holography/holoscopy setup.

5.4.4 The Ewald's sphere in holoscopy

As shown in Section 2.5.3 the measured angular spectrum of the object field in scattering theory with a plane incident field with wave vector k_0 and amplitude A_I is given by

$$\tilde{O}(k_x, k_y, z) = -A_I \frac{i}{2k_z} e^{+ik_z z} \tilde{\eta}(k_x - k_{x0}, k_y - k_{y0}, k_z - k_{z0}).$$

In holoscopy multiple wave vectors with varying lengths $\|k_0\|$ but the same incident direction are used with $k_{x0} = k_{y0} = 0$ and thus $k_{z0} = k_0 = k$. The amplitude is $A_C(k) A_O$, a function of wavenumber. The resulting object wave field is given by

$$\tilde{O}(k_x, k_y, k; z) = -\frac{i A_C(k) A_O}{2k_z} e^{+ik_z z} \tilde{\eta}(k_x, k_y, k_z - k). \tag{5.4.17}$$

Inverting (5.4.17) gives

$$A_C(k) \cdot \tilde{\eta}(k_x, k_y, k_z - k) = i \frac{2k_z}{A_O} e^{-ik_z z} \tilde{O}(k_x, k_y, k). \tag{5.4.18}$$

From this perspective it is clear, that only a subregion of the frequency domain representation of the scattering potential can be covered as shown in Figure 5.4.4. Especially the DC part of the potential is not obtained and therefore no absolute values of the refractive index of the sample can be determined.

It also becomes obvious, why the resampling step in the reconstruction is required: the data, as it is acquired in the frequency domain, is not acquired in the

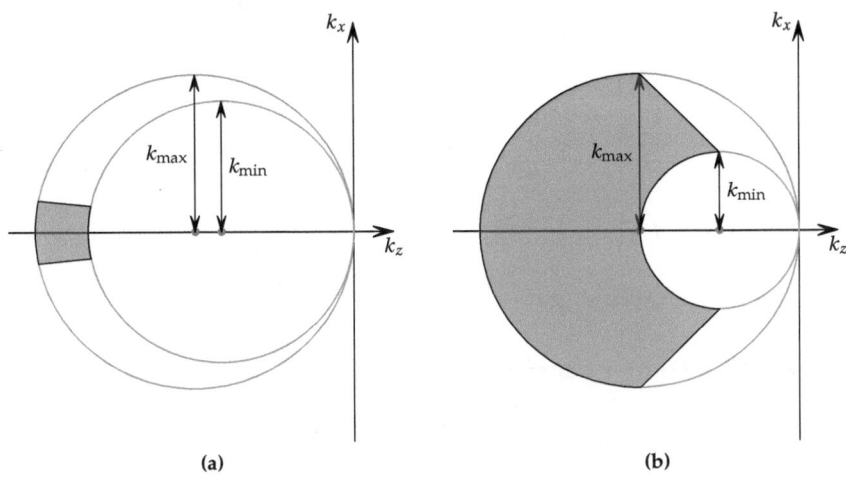

Figure 5.4.4.: Frequency components of the scattering potential that can be obtained by holoscopy for a sweep-range between minimal and maximal wavenumber k_{min} and k_{max}, respectively. The lateral frequencies are limited by the NA. The blue areas mark the Fourier components of the scattering potential, that can be obtained by using holoscopy. a) For low sweep-range at NA 0.2. b) For high sweep-range at NA 1.0.

Cartesian coordinate system of the frequency domain. The coordinate transformation shown in Figure 5.4.2 corrects for this problem.

5.4.4.1 Relation between scattering theory and one-step holoscopy reconstruction

In the derivation of the reconstruction equations for holoscopy in Section 5.3, and also for OCT in Section 2.7, the sign convention for the scattered and incident field in z-direction is different. The OCT cross-correlation term has a factor $\cos(2kz)$ instead of $\cos((k_i - k_s)z)$, where k_i is the incident wave vector and k_s is the backscattered wave vector and both are related by $k_s = -k_i = k$ in case of backscattering. Accordingly, the backscattering vector has different sign, i.e. k_z needs to be replaced by $-k_z$ to match the coordinate systems. Furthermore distances in the holoscopy derivation are measured positive, while for backscattering the plane of observation is behind the origin; this results in the camera plane being given as $z = -z_0$ when going from scattering theory to holoscopy. Finally, in scattering theory the obtained phase is determined by the phase of the incident wave. In the derivation of scattering theory, it was assumed that the incident wave has a zero phase in the origin $z = 0$. This gives the following translation table to go from scattering theory to holoscopy:

$$k_z \rightarrow -k_z$$
$$z \rightarrow -z_0$$
$$O \rightarrow e^{+ik z_0} O_0$$

This yields for (5.4.18)

$$A_C(k) \cdot \tilde{\eta}\left(k_x, k_y, -k_z - k\right) = -i\frac{2k_z}{A_O} e^{-ik_z z_0} e^{ikz_0} \tilde{O}_0\left(k_x, k_y, k\right).$$

Taking the inverse Fourier transform with respect to k on both sides gives

$$\tilde{A}_C(z) * \left(\tilde{\eta}(k_x, k_y, -z)e^{-ik_z z}\right) = -\frac{i}{2\pi}\int dk\, \frac{2k_z}{A_O} e^{-i(k_z - k)z_0} e^{ikz} \tilde{O}_0\left(k_x, k_y, k\right).$$

This can be rewritten to

$$\left(e^{ik_z z}\tilde{A}_C(z)\right) * \tilde{\eta}\left(k_x, k_y, -z\right) = -\frac{2i}{A_O}\int dk\, k_z e^{+i(k_z+k)z} e^{-i(k_z-k)z_0} \tilde{O}_0\left(k_x, k_y, k\right).$$

Finally, replacing z with $-z$, i.e. mirroring the coordinate system in the origin gives the formula

$$\left(e^{-ik_z z}\tilde{A}_C(-z)\right) * \tilde{\eta}\left(k_x, k_y, z\right) = -\frac{2i}{A_O}\int dk\, k_z e^{-i(k_z+k)z} e^{-i(k_z-k)z_0} \tilde{O}_0\left(k_x, k_y, k\right),$$

an expression similar to the one-step holoscopy reconstruction (5.4.2). The additional phase factor arises due to the different origins of the z coordinate system and will not change the amplitude of the results. The factor k_z indicates a different weighting of the angular components and should have minimal effect on the reconstruction process. The different factors arise due to the different approaches and the different mathematical formalism.

5.5 Setup considerations

5.5.1 Restrictions on the lens-less setup by the Nyquist criterion

In order to reconstruct holoscopic or holographic data completely, the interference fringes must be sampled correctly by the area camera. This gives certain restrictions on the setup geometry, which will be analyzed in this Section. Similar considerations for simpler setups, used in digital holography, are found in [52].

The local spatial frequencies of the occurring fringes depend on the local angles between reference and sample beam or – from an equivalent point of view – on the angle between reference and sample wavefront. Assuming two plane waves, a sample and a reference wave with wave vectors k_{obj} and k_{ref}, and amplitudes A_R and A_O, respectively. Superimposing them gives the intensity distribution in position space by

$$
\begin{aligned}
I(x) &= \gamma \left| A_R e^{i k_{ref} \cdot x} + A_O e^{i k_{obj} \cdot x} \right|^2 \\
&= \gamma \left[A_R^2 + A_O^2 + 2 A_R A_O \operatorname{Re}\left(e^{i(k_{ref} - k_{obj}) \cdot x} \right) \right].
\end{aligned}
$$

The fringe frequency is given by the difference of the k-vectors and their angle to the x-vector. As shown in Section 5.5.1, for a given sampling spacing Δx, the highest one-dimensional frequency K signal that can still be sampled is limited by the Nyquist criterion (5.5.1) to $K < \pi/\Delta x$. It follows the restriction for the holoscopy setup:

$$
\left\| k_{ref} - k_{obj} \right\| \lesssim \frac{\pi}{\sqrt{2}\Delta x}, \tag{5.5.1}
$$

for all beams hitting the camera. The factor $\sqrt{2}$ is considering the limiting case of sampling along the diagonal axis of the pixels.

One can restrict attention to a one-dimensional problem by choosing a suitable coordinate system, and let an arbitrary position on the camera be denoted by x and an arbitrary position on the sample be denoted by ℓ (Figure 5.5.1). These two variables will be restricted by the size of the sample or the size of the camera to

$$
x \in \left[-\sqrt{2}\frac{N\Delta x}{2}; +\sqrt{2}\frac{N\Delta x}{2} \right] \quad \text{and} \quad \ell \in \left[-\frac{L}{2}; +\frac{L}{2} \right],
$$

respectively, where N denotes the number of camera pixels of size Δx, and L is the diameter of the illuminated part of sample. One can compute the axial components of the wave vectors using simple geometry by applying the intercept theorem:

$$
\frac{k_{ref,x}}{k} = \frac{x}{\sqrt{(d-f)^2 + x^2}} \quad \text{and} \quad \frac{k_{obj,x}}{k} = \frac{x - \ell}{\sqrt{d^2 + (x - \ell)^2}}. \tag{5.5.2}
$$

By inserting (5.5.2) into (5.5.1), one obtains an inequality that needs to be true for

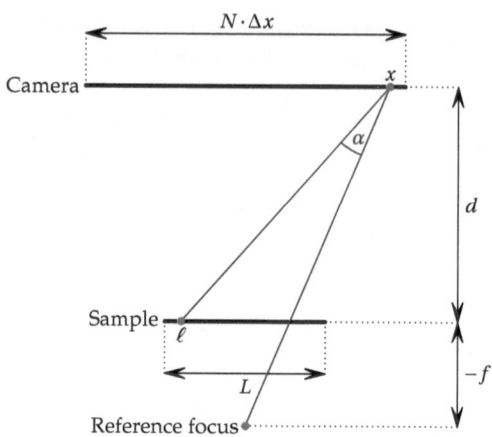

Figure 5.5.1.: For a lens-less holoscopy setup, spatial frequencies on the camera depend on the local angle α between reference and sample beam. By assuming plane waves, varying their origin on the sample ℓ, and their destination on the camera x one can probe whether all occurring spatial frequencies can be sampled correctly by the camera.

all pixels of the camera x and for all positions of the sample ℓ:

$$\left| k_{\text{ref},x} - k_{\text{obj},x} \right| = k \left| \frac{x}{\sqrt{(d-f)^2 + x^2}} - \frac{x-\ell}{\sqrt{d^2 + (x-\ell)^2}} \right| \lesssim \frac{\pi}{\sqrt{2}\Delta x} \qquad (5.5.3)$$

Failing to satisfy this condition will cause signals originating from the sample to be sampled incorrectly – at least on parts of the camera. Ghost images, caused by aliased signals, and artifacts will show up in the reconstructed tomograms. The condition (5.5.3) can be evaluated numerically, since the computation is quite fast and optimizations can be performed in reasonable time by simple parameter searching.

Restrictions for a concrete setup   To analyze the restrictions implied by (5.5.3), various test scenarios were considered. First, a setup as shown in Figure 5.2.1a was evaluated for parameters closely matching the actually used components (see Section 5.6.3). The experimental parameters were fixed, such as the size of the object $L = 3$ mm, the pixel pitch $\Delta x = 8\,\mu$m, the number of pixels $N = 1024$, and the focal length of the reference mirror $f = -10.34$ mm. For these parameters, the minimal distance d was obtained and the resulting NA and lateral resolution was

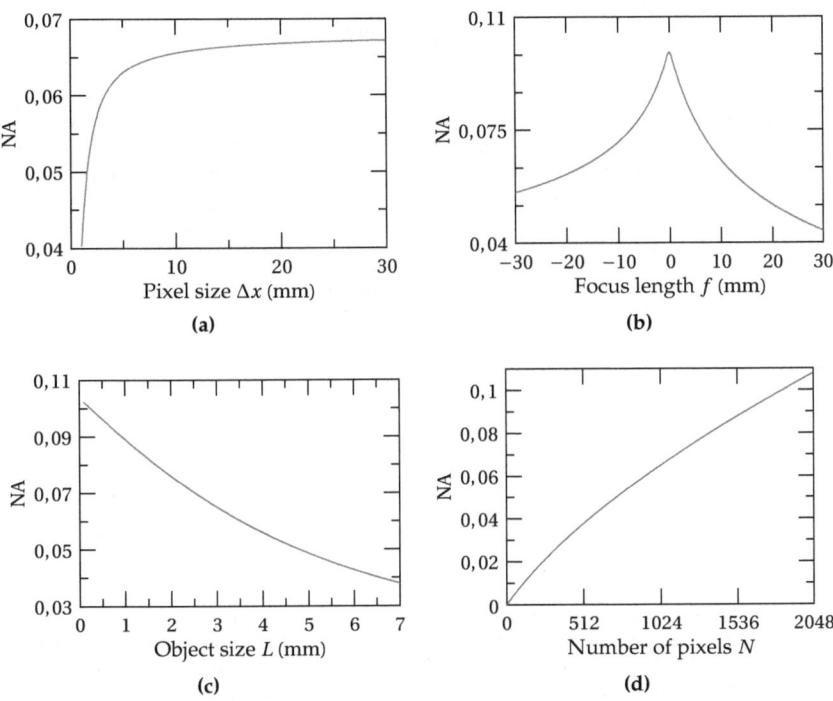

Figure 5.5.2.: Dependence of the possible maximal NA on the setup geometry shown in Figure
5.2.1 for $\lambda = 823.5$ nm. One parameter is varied, while the other parameters out of $L = 3$ mm,
$\Delta x = 8$ μm, $N = 1024$ and $f = -10.34$ mm are fixed according to the setup used (see
Section 5.2). For each set of parameters the equation (5.5.3) needs to be fulfilled for all
$\ell \in [-L/2; +L/2]$ and $x \in [-N\Delta x/2; +N\Delta x/2]$. While these conditions are fulfilled, the distance
d between camera and sample was minimized numerically and the resulting maximum NA
was computed. a) For this setup the maximal NA increases with the pixel size. b) It is
optimal to let the reference wave originate from within the sample. c) The maximum possible
NA decreases with increased sample size. d) The NA can be increased when using more
pixels in total, however in this case numerical complexity also increases significantly (see
Section 5.6.1.1).

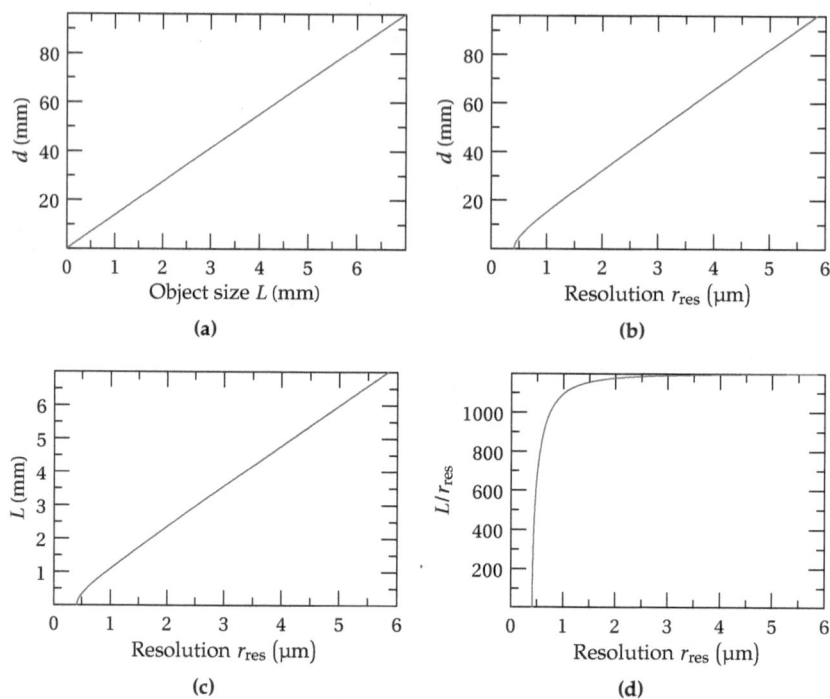

Figure 5.5.3.: Assuming a setup that is not restricted by any geometrical considerations, and setting the reference wave origin to the reference plane, i.e. $f = 0$, the resolvable NA is limited by the achievable field-of view and the camera. However, distances between sample and camera become arbitrarily small. This makes an experimental implementation impossible.

computed. Finally, one of the four parameters was taken, varied and the resulting maximal imaging NA was obtained. The results are shown in Figure 5.5.2.

Fundamental restrictions   The so far stated restrictions on the setup are not fundamental. In principle, the spherical reference wave can originate from the sample plane and thus $f = 0$ results in an optimal NA. As shown in Figure 5.5.3, the NA can become very large in this scenario, as does the resolving power of the holoscopy setup. To achieve this, the sample and the origin of the reference wave need to be very close to the camera plane, but this imposes significant problems. The space between camera and sample becomes too small to contain beam splitters or other optical devices and the reference wave can therefore not be superimposed on the sample wave. This scenario is therefore difficult or impossible to implement.

5.5.2 Considerations with imaging optics

Higher resolutions can be achieved by using additional high NA optics. In this Section basic considerations, when using imaging optics for holoscopy, are considered. In principle these considerations also need to be taken into account, when dealing with magnification in lens-less holoscopy. However, for low NA they are significantly less severe.

5.5.2.1 Reconstruction with a simulated lens

In contrast to lens-less holoscopy the imaging optics needs to be taken into account for a high-resolution holoscopy setup (Figure 5.2.1b). In general, imaging optics will cause the required reconstruction distance to depend non-linearly on the optical path lengths. The multiplication with a spherical reference wave with radius of curvature R is equivalent to a numerical lens with a focal length of $f_{rec} = R$ (see the representation of thin lenses in Section 2.4.3, especially (2.4.7) and the paraxial approximation of the reference wave (5.3.7)). As shown in Figure 5.5.4, the required reconstruction distance z_{rec} is obtained by the imaging equations

$$\frac{1}{z_{obj}} + \frac{1}{z'} = \frac{1}{f} \quad \text{and} \quad \frac{1}{d - z'} + \frac{1}{z_{rec}} = \frac{1}{R}.$$

Here z_{obj} denotes the distance of the object layer to the principle plane of the microscope objective, z' is the position of the intermediate image relative to the objective, d the propagation distance from microscope objective to camera and R the radius of the reconstruction wave, which is equivalent to the focal length of the reconstruction lens. Eliminating z', this yields

$$z_{rec} = \frac{(f - d)Rz_{obj} + dfR}{(f - d + R)z_{obj} + fd - fR}. \tag{5.5.4}$$

The magnification of the reconstruction is then given by

$$M = M_1 M_2$$

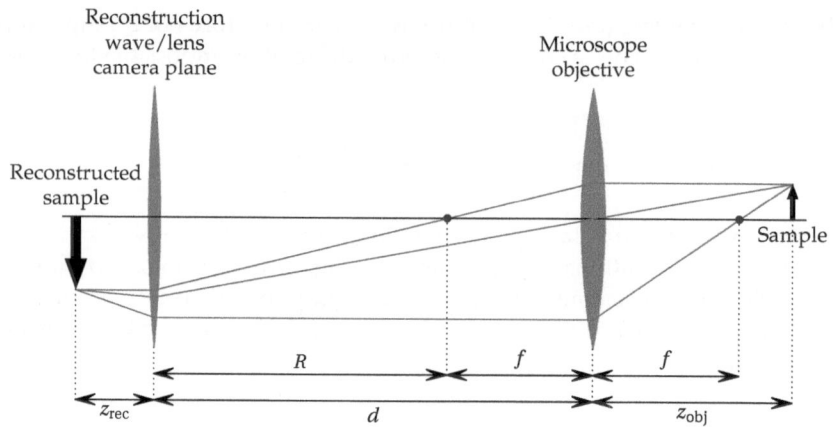

Figure 5.5.4.: Reconstruction of holoscopic data using a microscope objective. A spherical lens and a spherical wave have the same effect on the data. The spherical wave therefore serves as imaging lens.

with

$$M_1 = -\frac{z'}{z_{obj}} = -\frac{f}{\left(z_{obj} - f\right)} \quad \text{and} \quad M_2 = -\frac{z_{rec}}{d - z'} = -\frac{z_{rec}}{d - \left(f z_{obj}\right)/\left(z_{obj} - f\right)}.$$

The non-linearity between object distance z_{obj} and reconstruction distance z_{rec} in (5.5.4) poses a problem. Although in principle, it can be taken into account during reconstruction, the approaches described in Section 5.4 cannot be applied one-to-one. Additionally, if part of the sample is positioned in the focus of the imaging system composed of microscope objective and the lens simulating the reference wave, the reconstruction distance will diverge at this point making a general reconstruction impossible. For simple holographic reconstruction of a single layer this does in general not pose a problem. In digital holography these effects are only taken into account when objects need to be tracked as the required refocusing distance and the phase change of the signal are no longer identical [112–114].

Alternatively, the reconstruction distance (5.5.4) becomes a linear relation of the object layer distance z_{obj}, if

$$R = d - f. \tag{5.5.5}$$

Thus the reconstruction radius is restricted by (5.5.5), i.e. the numerical computations need to correspond to a $4f$-setup. The resulting formula is then given by

$$z_{rec} = -\frac{(d-f)^2}{f^2}z_{obj} + \frac{d(d-f)}{f} = -\frac{d-f}{f}\left(\frac{d-f}{f}z_{obj} - d\right).$$

The overall magnification in this case is given by

$$M_1 M_2 = \frac{f z_{rec}}{d\left(z_{obj} - f\right) - f z_{obj}} = -\frac{d-f}{f}.$$

5.5.2.2 Lens-less reconstruction

An equivalent point-of-view is achieved by transforming the imaging system to a lens-less setup. As shown in Figure 5.5.5, one assumes the camera is imaged by the microscope objective into a plane close to the sample. The size of the virtual camera, including the pixel spacing, is reduced by this approach. The sample is then imaged lens-less onto this "virtual" camera. Assuming, we have a distance d from camera to the principal plane of the objective with focal length f, the distance of the "virtual" camera to the principal plane d' is given by

$$\frac{1}{d'} + \frac{1}{d} = \frac{1}{f}.$$

The according magnification of the virtual camera is $M = -\frac{d'}{d}$ and the new pixel size to be used for computations is given by

$$\Delta x' = \left|-\frac{d'}{d}\right|\Delta x = |M|\Delta x.$$

For this approach the reference wave needs to be adapted accordingly. In the physical setup (Figure 5.2.1b), the physical reference wave is plane, when illuminating the real camera. But the reference wave needs to be transformed in the same way the object wave is transformed, when assuming an illumination on the virtual camera. For this, one assumes a microscope objective, identical to the one used in the sample arm and in the same position relative to the camera, has been placed in the reference arm to achieve a collimated beam. A spherical reference wave originating at the focal plane of the objective is required to achieve the collimated reference beam. As the camera has a distance d' from the principle plane of the objective and the reference wave an origin at the focus f, the reference will have a radius of curvature R of

$$R = d' - f$$

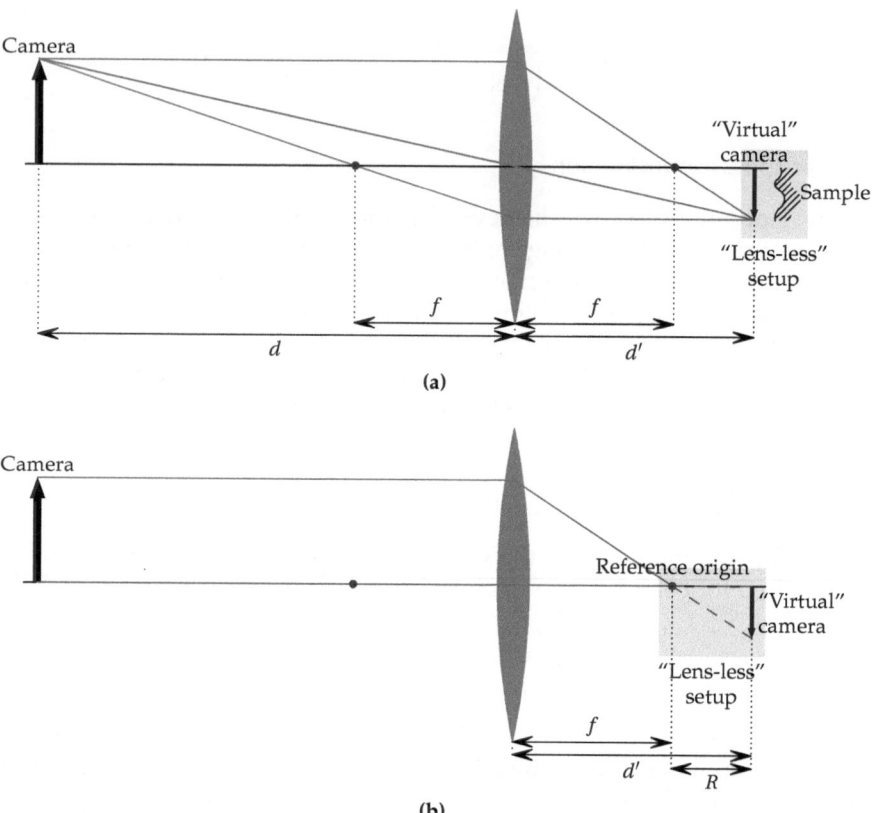

Figure 5.5.5.: Imaging optics through a microscope objective. By transforming camera position, camera size (a), reference wave origin, and curvature (b), the acquired data can be assumed to come from a lens-less setup.

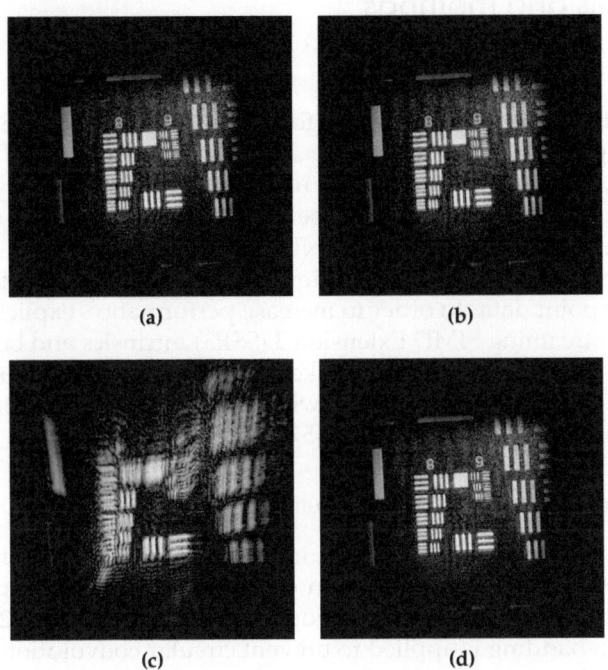

Figure 5.5.6.: Holograms of a US Air Force test chart acquired with a 40× magnification microscope objective (W N-Achroplan 40×/0.75, Zeiss) and water immersion at 867 nm. a) Non-
paraxial angular spectrum reconstruction with a numerical lens (Section 5.5.2.1). b) Paraxial
reconstruction with a numerical lens (Section 5.5.2.1) c) Non-paraxial angular spectrum
lens-less reconstruction (Section 5.5.2.2) d) Paraxial lens-less reconstruction (Section 5.5.2.2).

at the "virtual" camera plane. The transformation of the camera to a lens-less setup
is completely equivalent to the imaging optics point-of-view.

Although the lens-less point of view allows a setup using imaging optics to be
treated in the same way as a lens-less setup, it has some problems. While the wave
fields were physically captured using a pixel size that was significantly larger than
the wavelength, the virtual camera has virtual pixels that are much smaller and
closer to the wavelength. As the highest possible occurring interference fringes
have a spacing of half a wavelength, it becomes obvious that this transformation to
the lens-less point of view does not describe the situation perfectly. However, computations can still be performed, if the paraxial approximation is used (Figure 5.5.6).

5.6 Materials and methods

5.6.1 Implementation

For implementation the reconstruction formulas need to be discretized, which is described in Appendix C.

The code was implemented in C++ using the standard library according to ISO C++01. A minimalistic graphic user interface based on the Qt framework was created. As compiler g++ of the GNU Compiler Collection (GCC) was used for the x64 architecture. All computations were performed on single-precision (32 bit) floating point data. In order to increase performance, explicit vectorization using the Intel Streaming SIMD Extension 3 (SSE3) intrinsics and below, as well as OpenMP were used. Care needs to be taken to allow optimal memory usage, most computations were performed in-place with data in memory. To compute the FFT, the FFTW library version 3.2 was used [57].

5.6.1.1 Numerical complexity and execution speed

The simple reconstruction by (5.4.1) requires for one focal volume the propagation of each acquired hologram to a certain depth. For a single propagation of an image array of size $N_X \times N_Y$ two 2D Fourier transforms of size $2N_X \times 2N_Y$ are required, if zero-padding is applied to prevent circular convolution artifacts – one transform to go from position space to Fourier space and one to go back. Consequently, for N holograms at different wavelengths a total of $2N$ two-dimensional Fourier transforms of size $2N_X \times 2N_Y$ are required. For each lateral position a one-dimensional Fourier transform of size N needs to be performed afterwards to gain depth information, i.e. additional $N_X \times N_Y$ Fourier transforms calculate the final data. In total, $2N$ Fourier transforms of $2N_X \times 2N_Y$ arrays plus $N_X \times N_Y$ one-dimensional Fourier transforms of N data points are needed for each focal volume. The overall time complexity C_{SL} of a simple reconstruction with a single focus layer is of the order of

$$C_{SL} \sim \mathcal{O}(8NN_XN_Y \cdot \log(4N_XN_Y) + NN_XN_Y \cdot \log N).$$

One can assume, there are N_F simple reconstructions required to get all depth layers of the volume reconstructed sharply, i.e. the ratio between measurement depth and confocal parameter ($2z_R$) is N_F. In this case the sharp reconstruction of the entire volume, comprising multiple layers, has overall time complexity C_{ML} given by

$$C_{ML} \sim \mathcal{O}(8NN_FN_XN_Y \cdot \log(4N_XN_Y) + NN_FN_XN_Y \cdot \log N).$$

The one-step reconstruction of the complete volume by (5.4.6) reduces the computational complexity significantly. It also requires N two-dimensional Fourier transforms of size $2N_X \times 2N_Y$ to bring the images to Fourier space and $2N_X \times 2N_Y$ one-dimensional Fourier transforms of size N to gain depth information. For the complete volume this is $4\times$ more than for the reconstruction of only one focal volume by propagation and FFT. Because of the hermitian symmetry, the inverse transform from 2D Fourier space to position space only requires $N/2 + 1$ two-dimensional Fourier transforms of size $2N_X \times 2N_Y$, which is about half the amount required for the propagation and FFT approach. The overall time complexity C_{FV} of the one-step reconstruction for the full volume can thus be written as

$$C_{FV} \sim \mathcal{O}(6NN_XN_Y \cdot \log(4N_XN_Y) + 4NN_XN_Y \cdot \log N).$$

On a quad-CPU Opteron 6150 a reconstruction of a dataset of 1024 holograms with 1024×1024 pixel by (5.4.1) for one focal volume took about 22 s whereas a reconstruction of the complete volume by (5.4.2) or (5.4.6) took about 40 s, i.e. about twice the time required for the reconstruction of the focal volume. For a lens-less setup with 0.05 NA, the confocal parameter (i.e. twice the Rayleigh length) was $2z_R = 220\,\mu m$ and the measurement depth was 3.7 mm. Thus a data set would need 17 reconstructions of different focal volume for an overall diffraction limited resolution in air. Hence the one-step reconstruction of the complete volume offered a speedup of about $8.5\times$ for the low NA setup (Figure 5.2.1a). For a medium with refractive index $n = 1.5$ – as the scattering sample in Figure 5.7.1 – this speedup is reduced by a factor $\zeta = 1/n^2 \approx 0.44$, because the focal range is increased by a factor n and the effective total measurement depth is the optical depth which is reduced by a factor $1/n$. The full reconstruction is still about $3\times$ to $4\times$ faster in this case.

For the high NA measurements shown in Figure 5.7.4, acquired using the Mach-Zehnder type setup (Figure 5.2.1b), the confocal parameter was reduced to about $2z_R = 30\,\mu m$ and the measurement depth was about 2.2 mm. Therefore in air about 70 reconstructions are required with (5.4.1) and an approximately $30\times$ total speed-up is achieved. With a refractive index of about $n = 1.4$ this speed-up is again reduced by a factor $\zeta \approx 1/2$ and thus the grape shown in Figure 5.7.4 can be reconstructed about $15\times$ faster using the complete reconstruction by (5.4.6) compared to the single layer reconstruction by (5.4.1).

For higher lateral resolution, this factor will increase further. The actual gain of time for the reconstruction depends on the ratio of confocal parameter $2z_R$ to the measurement depth d of the system and the refractive index n of the sample. The expected improvement in reconstruction speed is shown in Figure 5.6.1. At high NAs near unity the one-step reconstruction is therefore expected to be about three

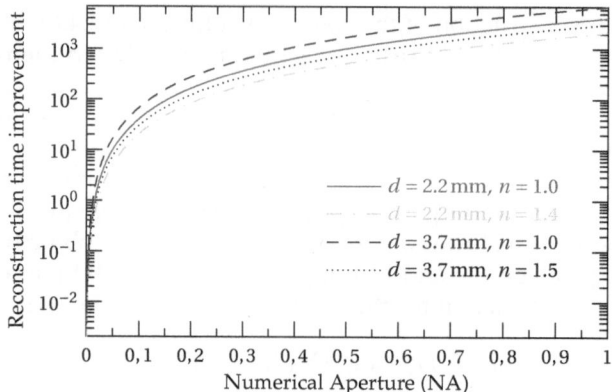

Figure 5.6.1.: Approximate increase of the reconstruction speed by using the one-step algorithm of (5.4.6) instead of sequentially applying (5.4.1) for multiple numerical foci. The increase of speed depends on the NA, the measurement depth d. and the refractive index n of the sample.

orders of magnitude faster.

5.6.2 Calibration

Dispersion and chirp   The used tunable-lasers are in general not sweeping linearly in the wavenumber k and additionally, for a high-resolution setup using a Mach-Zehnder interferometer, reference and sample arm are usually not dispersion matched. To compensate for this a single calibration measurement using a cover plate was performed. A chirp vector describing the wavenumber at a specific sweep time and a dispersion vector were computed from the data (Section 3.2).

Reference radius   For the lens-less setup (Figure 5.2.1a) the exact reference radius in the camera plane needs to be known to allow optimal imaging. In order to achieve this, a single hologram of a reflecting surface in the sample arm was acquired, using collimated sample illumination. By reconstructing with a plane reference wave and propagating the resulting interference pattern back to its focus, the exact reference radius was calculated.

Reconstruction distance   The distance between objective lens and camera, shown as d in Figure 5.5.4 and Figure 5.5.5, is not exactly known in the high-resolution setup. By propagating the object field from a single hologram with a known wave-

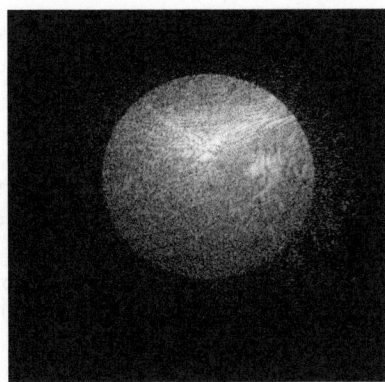

Figure 5.6.2.: Distance between camera and the plane of the objective is found by refocusing an acquired hologram without applying a reference wave, until the circular aperture is visible. The distance is found as the required propagation distance. The image shows the aperture, visible in a hologram of a coin, that was acquired using a Mach-Zehnder setup (Figure 5.2.1b).

length, back in the direction of the sample, the entrance pupil can be determined as a clearly visible aperture. The required propagation distance is then the distance from the principle plane of the objective to the camera.

5.6.3 Setup

5.6.3.1 Lens-less holoscopy

For the lens-less setup (Figure 5.2.1a) a Broadsweeper BS-840-1 (Superlum, Ireland) was used as swept-source laser. It provided a tuning range from 823.5 nm to 873.5 nm and 3 mW output power and according to (5.4.16) the FWHM of the axial PSF (axial resolution) was thus given by $\sim$15 µm.

A convex reference mirror with a focal length of about $f = -10$ mm was used for generating a spherical reference wave. In order to reduce reference light only a 4 % reflex of an air-glass boundary was used as reference mirror. In first experiments, a plan-convex uncoated lens with the plan side inked black was used. In this case the beam splitter cube had to be tilted to prevent its internal reflections to reach the camera. In later experiments an uncoated plan-concave lens has been directly attached to a beam splitter cube by optical contact bonding [115].

For *ex vivo* measurements a Mikrotron EoSens MC3010 CMOS camera with 1696 × 1710 pixels of size 8 µm × 8 µm each was used. With an effective speed of 440 frames/s at an area of interest of 1024 × 1024 pixels, 1024 frames were acquired.

The sample and the reference mirror were illuminated with a beam diameter of 2.2 mm in this case.

For *in vivo* demonstrations a high-speed Photron SA5 camera was used, that provided 7000 frames/s at full resolution of 1024×1024 pixel. Each pixel had a size of $20\,\mu m \times 20\,\mu m$. 1024 frames have been acquired during the wavelength sweep for each volume. An illumination diameter of 3.6 mm was used here.

Both cameras were hardware-synchronized to the swept-source laser, i.e. the first frame started the sweep of the laser.

By taking the restrictions on the setup geometry demonstrated in Section 5.5.1 into account, the distance from sample to camera was 8.0 cm for the *ex vivo* setup and 14 cm for the *in vivo* setup. The resulting NAs were thus approximately 0.05 and 0.07, respectively. By taking the rectangular aperture into account, the respective lateral resolutions are computed as shown in Appendix B.2.2 to about $9\,\mu m$ and $6\,\mu m$.

5.6.3.2 High-resolution holoscopy

For the experimental verification of high-resolution holoscopy, the light of a newer model of the Superlum Broadsweeper BS-840-1, with a sweeping range from 800 nm to 882 nm was used. As camera a Basler ACE acA2040-180km was used, that can acquire frames of 2048×2048 pixels at 180 Hz frame-rate. However, due to the possible sweep-rates provided by the electronics of the Broadsweeper, the camera speed needed to be reduced to about 125 Hz to match the sweep-rate. The collimated reference beam had a diameter of 16 mm. For first experiments a $5\times$ Mitutoyo Plan-IR microscope objective (MO) was used. The collimator in the sample arm had a beam diameter of 2.2 mm and illuminated the sample through an $f = 75$ mm lens and through the MO. Polarization filters in front of the camera and after the collimator in the sample arm were used in combination with a quarter-wave plate between microscope objective and sample to reduce parasitic reflexes from the objective and the beam splitter.

5.7 Results

5.7.1 Lens-less holoscopy

Holoscopic imaging was demonstrated using a scattering sample which contains $300 - 800$ nm sized iron oxide nanoparticles embedded in polyurethane resin [116]. Both proposed reconstruction algorithms were evaluated: first, the simple reconstruction according to (5.4.1), which first propagates the object field from the camera plane to one depth in the sample and then applies the Fourier transform

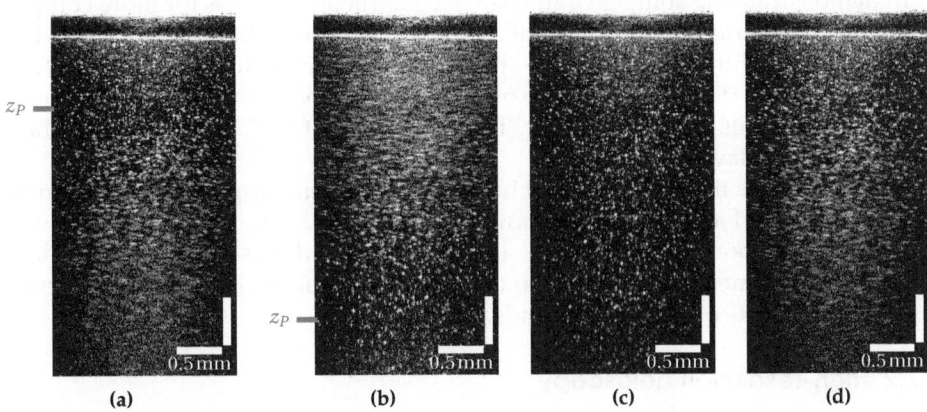

Figure 5.7.1.: B-scans from a reconstructed volume of a scattering phantom [116] consisting of multiple point scatterers. a) and b) result from reconstructions of the focal volume according to (5.4.1) at two different propagation depths z_P, which correspond to virtual numerical foci of the reconstruction. Outside the focal regions the lateral resolution is degraded. The confocal parameter was 220 µm. c) One-step reconstruction of the complete volume by (5.4.6) with the correct refractive index $n = 1.5$ ($\zeta = 0.44$). No lateral resolution degradation is visible. The loss of intensity in depth is caused only by a sensitivity roll-off due to the limited instantaneous coherence length of the laser source. d) One-step reconstruction of the complete volume by (5.4.2) without correcting for the increased index of refraction in the sample volume (i.e. $n = 1.0$ and thus $\zeta = 1$). Focus degradation is worse than in the reconstruction for a single focal volume. This is due to the fact that the former corresponds to $\zeta = 1$ and the latter to $\zeta = 0$. The correct value of $\zeta = 0.44$ is thus closer to the reconstruction of a single plane by (5.4.1).

with respect to the wavenumbers, yielded sharp tomograms only in the focal range around the reconstruction depth z_P (Figure 5.7.1a and b). This approach can be used, if only a single layer of the sample is of interest. The one-step reconstruction by (5.4.6) reconstructed the entire volume sharply; all layers over a depth of more than 30 Rayleigh lengths were obtained with diffraction limited resolution (Figure 5.7.1c). However, this only works if the index of refraction in the sample volume is correctly incorporated. A one-step reconstruction of a complete volume for the free space situation by (5.4.2) reconstructed only a limited depth region sharply (Figure 5.7.1d). In fact, with the phantom, that has a refractive index of about $n = 1.5$, the simple reconstruction shows better performance than the one-step reconstruction with $n = 1.0$, as the actual $\zeta = 1/n^2 = 0.44$ is closer to 0 than to 1.0. Only the full reconstruction by (5.4.6) with the correct ζ reconstructs all depths sharply.

To demonstrate the abilities of the new reconstruction process for more complex biological structures, a tomogram of a bug was acquired. In Figure 5.7.2 en-face images at three different depths are shown. The image quality of structures from within the bug is degraded because of refraction on the outer shell, caused by its non-homogeneous refractive index. The outer shell of the bug however is sharp within all depth-layers.

To demonstrate the capabilities of holoscopy for *in vivo* applications a finger-tip has been acquired and results are shown in Figure 5.7.3. The resulting acquisition rate here corresponds to about 7×10^6 A-scans/s and it is thus comparable to the fastest OCT measurements up to date. Imaging quality is sufficient to clearly visualize the ducts of the sweat glands.

5.7.2 High-resolution holoscopy

The advantage of holoscopy becomes more significant for higher NA since the Rayleigh length drops quadratically with the NA. Images of a grape were acquired with 0.14 NA using a microscope objective and a Mach-Zehnder interferometer. B-scans from reconstructed volumes clearly demonstrate that also for more complex structures a one-step reconstruction by (5.4.6) is possible while maintaining lateral resolution over the depth (Figure 5.7.4). Some additional artifacts were introduced by reflections of the microscope objective.

5.7.3 Artifacts in holoscopic imaging

In holoscopic imaging numerous artifacts occur, either because of additional light reaching the detector which cannot be filtered properly, or because the reconstruction process is not capable to re-obtain the actual scattering structures. Because of the lack of a confocal gating, holoscopic imaging is more sensitive to imaging artifacts than the measurements of a scanning OCT system. While some artifacts occur for both, confocal and scanning OCT, they might appear differently. Others artifacts are restricted to either one of the two technologies.

5.7.3.1 Autocorrelation artifacts

The sample can have various path lengths that occur in the autocorrelation term, causing additional signals in the tomographic data as shown in the area of Figure 5.7.5, marked by (A). In off-axis holoscopy (Figure 5.2.1b), these artifacts can be filtered to some degree. In on-axis, lens-less holoscopy setups this is not possible, and the ratio between sample and reference light cannot be adjusted arbitrarily, resulting in more dominant autocorrelation artifacts. The autocorrelated signals will not be imaged sharply, making them appear mostly as an additional noise

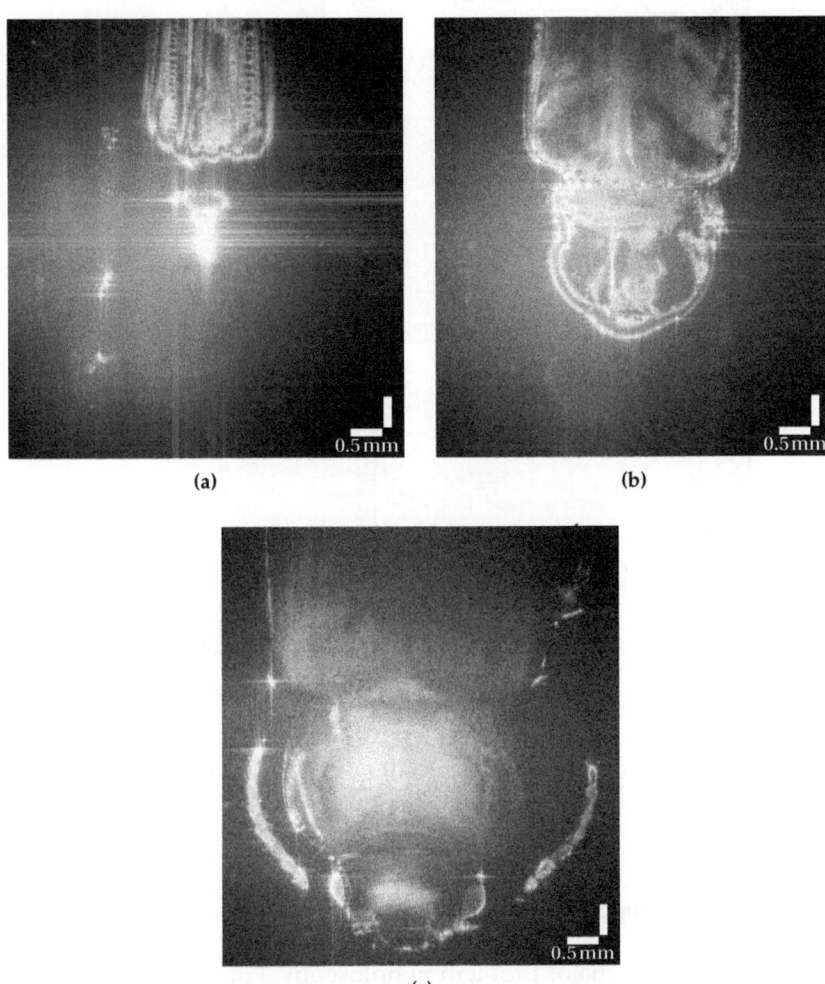

(a)   (b)

(c)

Figure 5.7.2.: En-face tomographic images of a bug at three different layers. The image cube was acquired by holoscopy. For reconstruction the one-step algorithm described by (5.4.6) was used. Internal structures of the bug can clearly be seen.

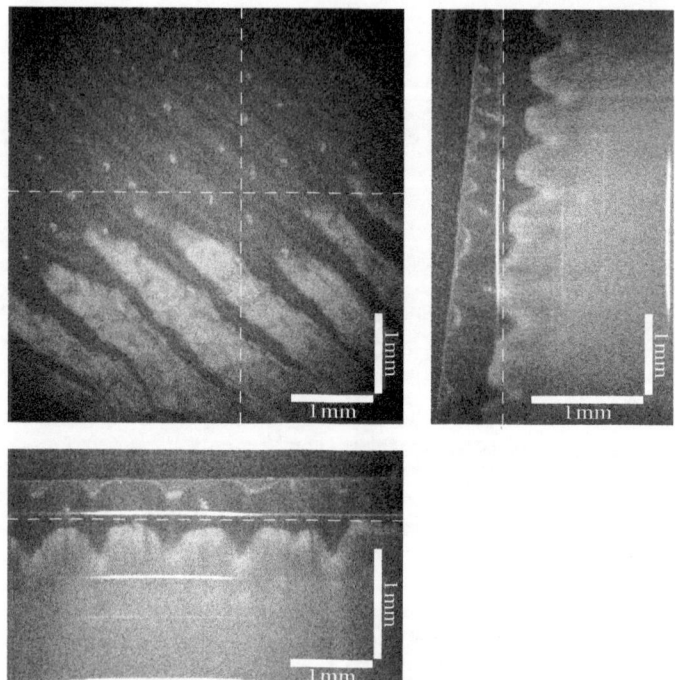

Figure 5.7.3.: Volume measurement of a finger tip acquired using holoscopy. The acquisition speed of 7000 frames/s corresponds to about 7×10^6 A-scans/s.

floor with speckles. This is in contrast to FD-OCT where the imaging structures can oftentimes be clearly identified as twin images.

5.7.3.2 Multiple scattering

Multiple scattering is a major problem in holoscopy. For the images shown, artifacts of multiply scattered photons are rather neglectable. In a single focus layer reconstruction as described in Section 5.4.1, the multiple scattered photons will in most scenarios be assigned to a false, higher depth, their lateral origin will be correct. This can be seen in Figure 5.7.5 in the area marked by (B) which shows a bug's antenna. Strongly scattering structures therefore appear to have a shadow-like vertical structure below them.

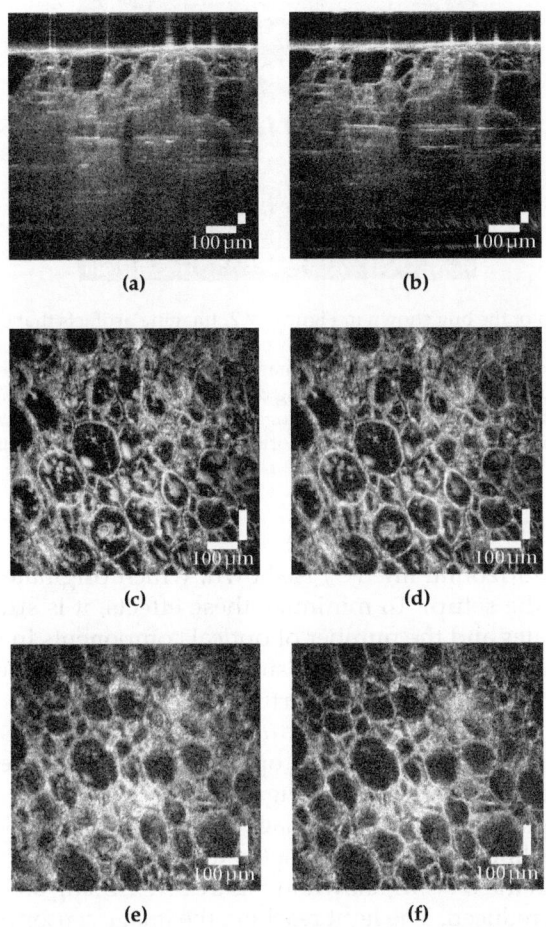

(a) (b)

(c) (d)

(e) (f)

Figure 5.7.4.: Holoscopic images of a grape acquired at 0.14 NA using the Mach-Zehnder type high-resolution setup. Simple reconstruction by propagating the field to one focal plane (left column) is compared with the one-step reconstruction of the complete volume by an NFFT (right column). a) B-scan of the simple reconstruction of a focal volume according to (5.4.1). b) B-scan of the one-step reconstruction according to (5.4.6). c) En-face image of the focal plane of the simple reconstruction. d) En-face image of the same plane in the one-step reconstruction. e) En-face image of the simple reconstruction in an optical distance of about 160 μm from the virtual focus shows deteriorated resolution. f) En-face image of a one-step reconstruction of the same layer. No degradation of the lateral resolution is observed. The confocal parameter was 28 μm. Remaining artifacts arise because of reflections from within the setup.

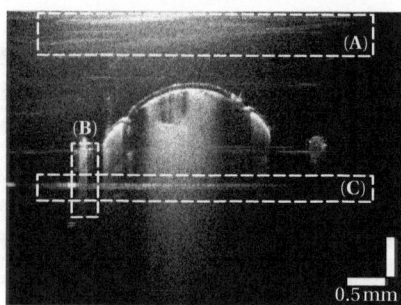

Figure 5.7.5.: B-scan of the bug shown in Figure 5.7.2. Imaging artifacts that occur in holoscopic imaging are shown. Autocorrelation signals create an increased noise floor in the upper part of the image (A). Contrary to FD-OCT imaging, the parts of the image appear not sharp. Multiple scattering of photons cause wrong depth assignment of the structures resulting in a bright shadow beneath strongly scattering surfaces (B). Reflecting or scattering parallel surfaces from within the setup can cause horizontal lines in the B-scan data (C).

5.7.3.3 Horizontal lines

In Figure 5.7.5 a horizontal line (C) is shown, which originates from reflecting interfaces within the setup. To minimize these effects, it is strongly advised to reduce parallel plates and the number of optical components in the setup. Tilting of beam splitter cubes or filters can also suppress the artifacts. Additionally, using an off-axis geometry suppresses these artifacts significantly.

Even when optical transitions are reduced, as for the lens-less setup by using optical contact bonding and preventing optical transitions, these artifacts did not entirely disappear as demonstrated in Figure 5.7.5.

Also the cover-glass in front of the image sensor causes additional two to three distinct horizontal lines in the reconstructed tomographic data. These horizontal lines can be seen in the B-scans shown in Figure 5.7.3. Using an off-axis geometry, they can partly be reduced. The light reaching the image sensor from the reference light causes the strongest artifacts and will not have the off-axis carrier frequency when mixed with itself; its signal can therefore easily be filtered. However, the sample light passing the cover glass, will mix with the reference light and still contain the off-axis carrier frequency; its signal can therefore not be filtered.

6 Conclusion

Optical tomographic techniques play a significant and growing role in medicine and sciences. In three parts of this thesis, solutions for problems and limitations of FD-OCT and full-field FD-OCT were found, implemented, and evaluated. The emphasis was on algorithmic solutions; while most of these problems and limitations can also be solved in hardware, as was already demonstrated, software solutions have unique and profound advantages. In the final Chapter, Holoscopy was introduced as new imaging modality.

6.1 FD-OCT signal processing using the NFFT

The performance of FD-OCT systems depends on the signal processing routines. One important issue is the resampling of the sampled spectrum onto a linear k-scale. Suboptimal processing can significantly reduce image quality or unnecessarily reduce processing speed. Viable alternatives to the standard interpolated FFT exist. Depending on the OCT system, its acquisition speed, its sensitivity, and the object to be imaged, an almost optimal algorithm can be selected. For the high-speed Thorlabs Hyperion OCT (127 kHz A-scan rate, 61 dB SNR) a linearly interpolated FFT without oversampling already provides almost ideal imaging quality. Online processing at more than 100 kHz A-scan rate was demonstrated. For the Thorlabs Callisto (1.2 kHz A-scan rate, 85 dB SNR), interpolation with oversampling is necessary to preserve the image quality. In most scenarios the non-equispaced fast Fourier transform (NFFT) with $2\times$ oversampling and a cut-off parameter of 2 provides an image quality which in all tested real scenarios was indistinguishable from the mathematical precise computations. With several tenth of kHz, the achieved processing rates were in most cases more than sufficient for online processing of the OCT data.

The use of a linear-k-spectrometer allowed for an acceptable imaging quality when only using a standard fast Fourier transform, and thus maximized data processing throughput. However, the use of a linear-k-spectrometer has its disadvantages. Electronic noise, undersampled signals, and other artifacts are no longer suppressed by the resampling step, and thus they provide lower imaging quality. Additionally, the linear-k-spectrometer, which compensated deviations of the k-sampling only partially, failed to provide the optimal image quality due to residual chirp. For optimal image quality resampling was still necessary.

6.2 Motion and dispersion

Phase errors caused by axial motion in swept-source OCT and GVD mismatch result in axial blurring and reduced imaging quality. Both can be corrected by phase multiplications to the analytical OCT signal prior to the Fourier transform. It was shown, that the correcting phase can be directly calculated from the measured OCT data. Cross-correlation of sub-bandwidth reconstructions (CCSBR) was developed as a new algorithm for determining wavenumber-dependent phase errors from the FD-OCT data. This algorithm does not depend on functional optimization, or a polynomial representation of the phase errors. This is of importance when applied to motion induced phase errors, which can in general not be described by a polynomial of low degree. The proposed algorithm is not iterative and does not require a stopping criterion.

Motion artifacts in full-field FD-OCT have been corrected and the proposed algorithm worked reliable when the artifacts were not too severe. Dispersion artifacts have been corrected, even when the dispersion mismatch was introduced by the sample itself, e.g. by the individual length of the eye bulb in retinal imaging. The imaging quality and effective axial resolution of high-resolution FD-OCT could be improved.

While providing an improved imaging quality in FD-OCT, the importance of the proposed algorithm is higher for motion artifacts in the full-field scenario. One of the major drawbacks of full-field SS-OCT is the increased vulnerability to motion artifacts, especially as high-speed cameras are expensive and provide only a limited area-of-interest at sufficient imaging speed, which is of special importance in *in vivo* imaging. Similar problems arise in holoscopy. The proposed algorithm can be ported there and should provide similar performance.

6.3 Holoscopy

One of the most important shortcomings of FD-OCT is the limited depth of focus. It reduces sensitivity and resolution in out-of-focus layers, and thereby prevents FD-OCT from reaching higher resolution while still maintaining its advantage compared to time-domain OCT. In Chapter 5, it was shown that a combination of full-field SS-OCT with digital holography can fully provide this advantage, even for imaging at higher NA. An almost depth independent resolution and sensitivity was achieved by using a lens-less setup for low NA, and using a setup with magnifying optics for high NA.

The improved imaging capabilities of holoscopy come at the cost of an increased computational complexity. Reconstruction was demonstrated by using a stack of holographic reconstructions, followed by standard FD-OCT signal processing, and repeated for depths every few Rayleigh lengths. It was shown, that a reconstruction can equivalently be performed in a single step by resampling in frequency space. Similar resampling techniques have previously been applied in inverse scattering for scanning FD-OCT (ISAM, interferometric synthetic aperture microscopy) [33–35,47] and have also been proposed for full-field FD-OCT [46,47]. Possible resampling algorithms in frequency space, combined with the actual Fourier transform were investigated for the OCT scenario in the first part of this thesis.

Using these techniques, first *ex vivo* and *in vivo* images could be presented. They stand out by an imaging quality that is for suitable samples comparable to scanning FD-OCT while providing a lateral resolution that is not governed and limited by the imaging optics. Additionally, the *in vivo* imaging speed corresponded to 7×10^6 Hz A-scan rate, one of the fastest OCT systems up to date. The advantage of holoscopy can be further expanded by going to higher NA.

6.4 Outlook

Holoscopy has great potential. For high NA, it can improve imaging the same way Fourier-domain OCT has improved upon time-domain OCT for low NA. The higher the lateral resolution, the more apparent becomes this advantage. It therefore appears natural to put the emphasis of holoscopic imaging into the microscopic regime, providing high-resolution structural imaging comparable to histology or high-resolution time-domain optical coherence microscopy. But the way to achieve this is not trivial. Spatially coherent, tunable light sources with sufficient tuning range and high instantaneous coherence length are not available. So far, reflected light, which is introduced by optical components, especially microscope objectives, reduces sensitivity and imaging quality. Additionally, microscope objectives can only be corrected optimally for a single imaging plane (see e.g. [117]), and are usually corrected for a limited wavelength range: Holoscopy, on the other hand, does tomographic, three-dimensional imaging and requires a multitude of wavelengths.

Holoscopy also allows for numerical aberration compensation entirely in the computer by introducing customized numerical lenses; similar approaches have been shown in digital holography (see e.g. [118–120]). In addition to the aberration correction for microscope objectives when used out of focus, this is of special importance for sample induced aberrations. In high-resolution retinal imaging holoscopy could compensate for the imaging errors of the human eye. Holoscopy might prove to be an alternative to the combination of wave-front sensors, deformable mirrors, and scanning OCT.

Because of its massive parallelization, with up to several millions of A-scans acquired at once, it also enables higher imaging speed. The lack of moving parts is an advantage compared to scanning OCT. Still, it is more sensitive to motion artifacts and high-speed cameras need to be used.

For holoscopy to become widely accepted and usable in many areas, quite a few challenges need to be faced. Imaging artifacts are numerous. They need to be better understood and reduced by a combination of hardware changes, improvements, and suitable software algorithms. For example, motion artifact compensation, as was demonstrated for full-field FD-OCT, needs to be introduced to holoscopy. Multiple scattering remains a huge problem in holoscopic imaging; while the lack of a confocal gating allows imaging of out-of-focus layers with optimal sensitivity, it does not suppress stray light, parasitic reflexes and multiple scattered light as efficient as confocal imaging.

A Mathematical supplements

A.1 Definitions

Below is a list of definitions for function that are frequently used in this thesis.

- Rectangular function

$$\text{rect}(x) = \begin{cases} 1 & \text{for } |x| \leq \frac{1}{2} \\ 0 & \text{otherwise} \end{cases} \tag{A.1.1}$$

- Circular function in Cartesian coordinates

$$\text{circ}(x, y) = \text{rect}\left(\frac{1}{2}\left(x^2 + y^2\right)\right) = \begin{cases} 1 & \text{for } x^2 + y^2 \leq 1 \\ 0 & \text{otherwise} \end{cases} \tag{A.1.2}$$

- Circular function in spherical coordinates

$$\text{circ}(r) = \text{rect}\left(\frac{1}{2}r\right) = \begin{cases} 1 & \text{for } r \leq 1 \\ 0 & \text{otherwise} \end{cases}$$

o Triangle function

$$\triangle(x) = \begin{cases} 1 - |x| & \text{for } |x| \le 1 \\ 0 & \text{otherwise} \end{cases} \tag{A.1.3}$$

o Sinc function

$$\text{sinc}(x) = \frac{\sin(\pi x)}{\pi x} \tag{A.1.4}$$

o Dirac comb-distribution

$$\text{comb}(ax) = \sum_{n=-\infty}^{+\infty} \delta\left(x - \frac{n}{a}\right) \tag{A.1.5}$$

o Step function

$$\Theta(x) = \begin{cases} 1 & \text{for } x \ge 0 \\ 0 & \text{otherwise} \end{cases} \tag{A.1.6}$$

o Signum function

$$\text{sgn}(x) = \begin{cases} +1 & \text{for } x > 0 \\ 0 & \text{for } x = 0 \\ -1 & \text{for } x < 0 \end{cases} \tag{A.1.7}$$

o Hann window function with spatial width X

$$h_X(x) = \begin{cases} \frac{1}{2}\left(1 + \cos\left(\frac{2\pi}{X}x\right)\right) & \text{for } -X/2 \le x \le X/2 \\ 0 & \text{otherwise} \end{cases} \tag{A.1.8}$$

o Gaussian window function

$$g_\sigma(x) = \exp\left(-\frac{x^2}{2\sigma^2}\right)$$

o The dilated Kaiser-Bessel window [59,60] as used for the NFFT (Section 2.1.3.3) with oversampling α and cut-off parameter m

$$w(x) = \begin{cases} \frac{\sinh\left(b\sqrt{m^2 - \alpha^2 N^2 x^2}\right)}{\pi\sqrt{m^2 - \alpha^2 N^2 x^2}} & \text{for } |x| < \frac{m}{\alpha N} \\ \frac{\sin\left(b\sqrt{\alpha^2 N^2 x^2 - m^2}\right)}{\pi\sqrt{\alpha^2 N^2 x^2 - m^2}} & \text{otherwise,} \end{cases} \tag{A.1.9}$$

with the parameter b given by

$$b = \pi\left(2 - \frac{1}{\alpha}\right).$$

o Of special importance is also the convolution of two function $f(x)$ and $g(x)$ defined by

$$(f * g)(x) = \int dx'\, f(x')g(x - x').$$

A.2 Useful identities

o The Dirac comb-distribution can be represented as Fourier series

$$\text{comb}(ax) = \frac{1}{a} \sum_{n=-\infty}^{+\infty} e^{i2\pi nx/a}$$

o The triangle function $\triangle(x)$ is the convolution of the rectangular function $\text{rect}(x)$ with itself. This follows directly from (A.3.1) in combination with (A.3.10) and (A.3.9)

$$\triangle(x) = \text{rect}(x) * \text{rect}(x)$$

o The convolution is commutative and associative

$$f * g = g * f \quad \text{and} \quad (f * g) * h = f * (g * h)$$

A.3 The Fourier transform

A.3.1 Properties of the Fourier transform

o Linearity

$$\mathscr{F}[af(x) + bg(x)] = a\mathscr{F}[f(x)] + b\mathscr{F}[g(x)]$$

o Applying the Fourier transform twice corresponds to scaling and mirroring in the origin

$$\mathscr{F}[\mathscr{F}[f(x)]] = 2\pi f(-x)$$

o If $f(x)$ is real-valued, i.e. $f(x) \in \mathbb{R}$ its Fourier transform has Hermitian symmetry

$$\mathscr{F}[f(x)] \in \mathbb{R}, \quad \tilde{f}(k) = \tilde{f}^*(-k)$$

○ Fourier transform of a Hermitian symmetric function is a real function

$$f(x) = f^*(-x) \quad \Leftrightarrow \quad \tilde{f}(k) \in \mathbb{R}$$

$$f(x) \in \mathbb{R} \quad \text{and} \quad f(x) = f(-x) \quad \Leftrightarrow \quad \tilde{f}(k) \in \mathbb{R}$$

○ Convolution theorem

$$\mathcal{F}[f(x)] \cdot \mathcal{F}[g(x)] = \mathcal{F}[(f * g)(x)] = \mathcal{F}\left[\int dx' \, f(x')g(x' - x)\right] \quad \text{(A.3.1)}$$

○ Scaling of the argument

$$\mathcal{F}[f(ax)] = \frac{1}{|a|}\tilde{f}\left(\frac{k}{a}\right) \quad \text{(A.3.2)}$$

○ Shifting of the argument

$$\mathcal{F}^{-1}[\tilde{f}(k - k_0)] = f(x)e^{+ik_0 x} \quad \text{(A.3.3)}$$

○ Scaling of the Fourier kernel

$$\int dz \, f(z)e^{-i\alpha kz} = \tilde{f}(\alpha k)$$

○ Differentiation

$$\mathcal{F}\left[\frac{\partial f(x)}{\partial x}\right] = ik\tilde{f}(k) \quad \text{(A.3.4)}$$

○ Parseval's theorem

$$\int dx \, f(x)g^*(x) = \frac{1}{2\pi}\int dk \, \tilde{f}(k)\tilde{g}^*(k) \quad \text{(A.3.5)}$$

○ Plancherel's theorem

$$\int dx \, |f(x)|^2 = \frac{1}{2\pi}\int dk \, |\tilde{f}(k)|^2 \quad \text{(A.3.6)}$$

○ Fourier transform of a rotationally symmetric function $f(r, \phi) = f(r)$ in spherical coordinates, $x = r \cos \phi$ and $y = r \sin \phi$, i.e.

$$\tilde{f}(k_x, k_y) = \int d^2x\, f\left(\sqrt{x^2 + y^2}\right) \exp(-i(k_x x + k_y y)).$$

This gives the Fourier

$$
\begin{aligned}
\tilde{f}(k_r, k_\phi) &= \int_0^{2\pi} d\phi \int_0^\infty dr\, r\, f(r) \\
&\quad \times \exp\left(-i(k_r r(\cos k_\phi \cos \phi + \sin k_\phi \sin \phi))\right) \\
&= \int_0^{2\pi} d\phi \int_0^\infty dr\, r\, f(r) \exp\left(-i(k_r r \cos(\phi - k_\phi))\right).
\end{aligned}
$$

This yields

$$\tilde{f}(k_r, k_\phi) = 2\pi \int_0^\infty dr\, r\, f(r) J_0(k_r r) = \tilde{f}(k_r), \tag{A.3.7}$$

with J_0 being the Bessel function of the first kind

$$
\begin{aligned}
J_0(x) &= \frac{1}{2\pi} \int_0^{2\pi} d\phi \exp(-ix \cdot \cos \phi) \\
&= \frac{1}{2\pi} \int_0^{2\pi} d\phi \exp(-ix \cdot \cos(\phi - k_\phi)),
\end{aligned}
$$

for all k_ϕ due to rotational symmetry.

A.3.2 Important Fourier transforms

o Harmonic function

$$\mathscr{F}\left[e^{ik'x}\right] = \int dx\, e^{ik'x}\, e^{-ikx} = 2\pi\delta(k - k')$$

o Fourier transform of the δ-distribution

$$\mathscr{F}[\delta(x)] = 1$$

o The Gauss function with width $\sigma = 1$ is its own Fourier transform, except for a scaling factor

$$\mathscr{F}\left[\exp\left(-\frac{1}{2}x^2\right)\right] = \sqrt{2\pi}\exp\left(-\frac{1}{2}x^2\right)$$

○ Gaussian function

$$\mathscr{F}\left[e^{-ax^2}\right] = \sqrt{\frac{\pi}{a}}e^{-\frac{k^2}{4a}}$$

○ Spherical wave [53]

$$\mathscr{F}_{xy}^{-1}\left[\exp\left(-iz\sqrt{k^2 - k_x^2 - k_y^2}\right)\right] = \frac{ik}{z}\exp\left(-ik\sqrt{x^2 + y^2 + z^2}\right)$$

○ comb-distribution. Note, that (A.3.2) does not hold due to the convenience definition of the comb-distribution. As a special case, the comb-distribution is its own Fourier transform for a comb-spacing $\Delta x = \sqrt{2\pi}$.

$$\mathscr{F}\left[\sum_{n=-\infty}^{+\infty}\delta(x - n\Delta x)\right] = \mathscr{F}\left[\text{comb}\left(\frac{x}{\Delta x}\right)\right]$$

$$= \frac{2\pi}{\Delta x}\sum_{n=-\infty}^{+\infty}\delta\left(k - \frac{2\pi n}{\Delta x}\right) = \frac{2\pi}{\Delta x}\text{comb}\left(\frac{\Delta x}{2\pi}\cdot k\right) \quad (A.3.8)$$

○ Rectangular function

$$\mathscr{F}[\text{rect}(x)] = \frac{\sin(k/2)}{k/2} = \text{sinc}\left(\frac{k}{2\pi}\right) \quad (A.3.9)$$

○ Triangle function

$$\mathscr{F}[\Delta(x)] = \frac{\sin^2(k/2)}{(k/2)^2} = \text{sinc}^2\left(\frac{k}{2\pi}\right) \quad (A.3.10)$$

○ Signum function

$$\mathscr{F}[\text{sgn}(x)] = \frac{2}{ik}$$

A.3.3 The uncertainty relation of Fourier analysis

The L_2-norm for a complex-valued function f is defined as

$$\|f(x)\|^2 = \int dx\,|f(x)|^2.$$

The variance around zero gives a measure for the spread or localization of the function f. It is defined by

$$\sigma_x^2 = \frac{1}{\|f(x)\|^2} \int dx \, x^2 |f(x)|^2 = \frac{\|x \cdot f(x)\|^2}{\|f(x)\|^2} > 0. \qquad (A.3.11)$$

The variance of the same function in frequency space is accordingly given by

$$\sigma_k^2 = \frac{1}{\|\tilde{f}(k)\|^2} \int dk \, k^2 |\tilde{f}(k)|^2 = \frac{\|k \cdot \tilde{f}(k)\|^2}{\|\tilde{f}(k)\|^2} > 0. \qquad (A.3.12)$$

Without loss of generalization, we assume $f(x)$ is normalized

$$\int dx \, |f(x)|^2 = \|f(x)\|^2 = 1.$$

This can always be achieved by a suitable scaling factor, without changing the variance (A.3.11). It follows by Plancherel's theorem (A.3.6) the normalization of the according function in frequency space

$$\int dk \, |\tilde{f}(k)|^2 = \|\tilde{f}(k)\|^2 = 2\pi.$$

Using (A.3.4), the variance in frequency space can be expressed as

$$\begin{aligned} \sigma_k^2 &= \frac{1}{2\pi} \int dk \, |k\tilde{f}(k)|^2 \\ &= \frac{1}{2\pi} \int dk \, \left| \mathscr{F}\left[\frac{\partial f(x)}{\partial x} \right] \right|^2 \\ &= \int dx \, \left| \frac{\partial f(x)}{\partial x} \right|^2, \end{aligned} \qquad (A.3.13)$$

where the last equality follows again from Plancherel's theorem (A.3.6). The Schwarz inequality for the L_2-norm (see e.g. [63]) states

$$\int dx \, |g(x)|^2 \cdot \int dx' \, |h(x')|^2 \geq \left| \int dx \, g^*(x) h(x) \right|^2. \qquad (A.3.14)$$

By inserting $g(x) = xf(x)$ and $h(x) = \partial f(x)/\partial x$ in the left hand-side of (A.3.14) and by using (A.3.13) and (A.3.11), it follows

$$\int dx\, x^2 |f(x)|^2 \cdot \int dx' \left| \frac{\partial f(x)}{\partial x} \right|^2 = \sigma_x^2 \cdot \sigma_k^2. \tag{A.3.15}$$

We define the integral on the right hand-side of (A.3.14) to be equal to I and it follows

$$I = \int dx\, x f^*(x) \frac{\partial f(x)}{\partial x}.$$

Integrating by parts gives

$$I = \underbrace{x|f(x)|^2 \Big|_{-\infty}^{+\infty}}_{=0} - \underbrace{\int dx\, x f(x) \frac{\partial f^*(x)}{\partial x}}_{=I^*} - \underbrace{\int dx\, f^*(x) f(x)}_{=1}.$$

It follows $I + I^* = 2\,\mathrm{Re}\,I = -1$ and thus $|I| \geq 1/2$. Combining this results with (A.3.14) and (A.3.15) it follows

$$\sigma_x \sigma_k \geq \frac{1}{2}.$$

This results also generalizes for variances around other means. By simple shifting of the function f, its mean value can be adjusted.

A.4 Complex analysis

Complex analysis treats the topics of a complex valued functions of a complex argument. Of special interest are holomorphic functions. A function $f(z)$ is called holomorphic if it is complex differentiable. It can be shown that this condition is equivalent to fulfilling the Cauchy-Riemann partial differential equations. For a function

$$f(z) = f(x + iy) = u(x,y) + iv(x,y),$$

these conditions are given by

$$\frac{\partial u(x,y)}{\partial x} = \frac{\partial v(x,y)}{\partial y} \quad \text{and} \quad \frac{\partial u(x,y)}{\partial y} = -\frac{\partial v(x,y)}{\partial x}.$$

The exponential function, constants, sines, cosines, polynomials, as well as their compositions, additions, and multiplications are all holomorphic functions. The division of holomorphic functions is in general holomorphic, except at the resulting poles. Consequently, rational functions are holomorphic except on their poles. The real part $\text{Re}\, f(z)$, the imaginary part $\text{Im}\, f(z)$, or the complex conjugate $f^*(z)$ of a holomorphic function $f(z)$, on the other hand, is in general not holomorphic.

A holomorphic functions can be developed in a Laurent series, which can be seen as a generalization of the Taylor series, and is given by

$$f(z) = \sum_{n=-\infty}^{+\infty} a_n (z-c)^n,$$

with suitable a_n for any complex point c. The coefficients a_n are determined by Cauchy's integral formula

$$a_n = \frac{1}{2\pi i} \oint_\gamma dz\, \frac{f(z)}{(z-c)^{n+1}}$$

for any closed, counter-clockwise, single-winded curve γ around the point c. The values of $f(z)$ around a curve γ thus completely determine all function values of f in the area surrounded by γ.

For any given function and point c, the coefficient a_{-1} is called the residue of the function $f(z)$, usually given the notation

$$\text{Res}(f,c) = \frac{1}{2\pi i} \oint_\gamma dz\, f(z). \tag{A.4.1}$$

These residues can in many cases be determined quite easily and thus allow easy computation of the corresponding integral. For example, with $f(z)$ having a single pole at position c, the residue is given by

$$\text{Res}(f,c) = \lim_{z \to c}(z-c)f(z).$$

The existence of this limit also ensures that $f(z)$ has a single pole there.

A.4.1 Residual theorem

Given a function that is holomorphic in the area surrounded by γ except on a finite amount of points z_k, the contour integral along γ can be determined by

$$\oint_\gamma dz\, f(z) = 2\pi i \sum_{z_k} \text{Res}(f, z_k). \tag{A.4.2}$$

A.4.2 Jordan's lemma

Jordan's lemma states, for a function of the form

$$f(z) = e^{ikz} g(z)$$

with any $k > 0$ and any holomorphic $g(z)$, the contour integral on a contour $\gamma_r = re^{i\phi}, \phi \in [0; \pi)$ vanishes

$$\lim_{r \to \infty} \int_{\gamma_r} dz\, e^{ikz} g(z) = 0.$$

This helps in the evaluation of Fourier integrals. Having an addition curve $\gamma_\mathbb{R}$ along the real axis, the curves γ_r and $\gamma_\mathbb{R}$ can be combined to yield a closed curve γ. Only the integral along the real axis then contributes to the overall integral and is completely determined by the residuals enclosed by the curve γ.

B

Resolution in signal processing and optics

The width and form of signals in frequency space determine their width and form in position space and vice versa. This influences the axial point spread functions in OCT, given as Fourier transforms of the spectrum or the apodization window function, if spectral shaping is applied. It also influences the lateral point spread functions in classical imaging as Fourier transform of the aperture function. In this Section, the most important spectra/window functions and aperture functions, as well as their according point spread functions will be computed and analyzed.

For a more complete overview of resolution in signal processing and optics, see e.g. [49, 51, 106].

B.1 Apodization windows

B.1.1 Rectangular window

The rectangular window with width K is given by

$$w(k) = \text{rect}\left(\frac{k}{K}\right),$$

its inverse Fourier transform is

$$\mathscr{F}^{-1}\left[\text{rect}\left(\frac{k}{K}\right)\right] = \frac{K}{2\pi}\,\text{sinc}\left(\frac{K}{2\pi}x\right).$$

The full-width at half maximum (FWHM) of this signal is easily determined. With $\text{sinc}(0) = 1$, it follows for the half-width x_{FWHM} that

$$\text{sinc}\left(\frac{K}{2\pi}\cdot\frac{x_{\text{FWHM}}}{2}\right) = \frac{1}{2},$$

which numerically evaluates to

$$x_{\text{FWHM}} \approx 7.58\frac{1}{K}.$$

With a total signal width of K divided in N equal parts of spacing Δk, i.e. $K = N\Delta k$, the resulting spacing in Fourier space can be computed by (2.1.6). It follows for the FWHM

$$x_{\text{FWHM}} = N_{\text{FWHM}}\cdot\Delta x \approx 7.58\cdot\frac{1}{K} = 7.58\cdot\frac{1}{N\Delta k} = \frac{7.58}{2\pi}\Delta x.$$

Thus the number of pixels specifying the FWHM is about

$$N_{\text{FWHM}} \approx \frac{7.58}{2\pi} \approx 1.21.$$

B.1.2 Hann window

The Hann window $h_K(k)$ with spectral width K is defined by (A.1.8). It can be rewritten as

$$h_K(k) = \frac{1}{2}\left(1 + \cos\left(\frac{2\pi}{K}k\right)\right)\cdot\text{rect}\left(\frac{k}{K}\right).$$

Its inverse Fourier transform is most easily obtained by looking at the Fourier transform of all the terms individually

$$\mathscr{F}^{-1}\left[\frac{1}{2}\right] = \frac{1}{2}\delta(x)$$

$$\mathscr{F}^{-1}\left[\frac{1}{2}\cos\left(\frac{2\pi}{K}k\right)\right] = \frac{1}{4}\left(\delta\left(x - \frac{2\pi}{K}\right) + \delta\left(x + \frac{2\pi}{K}\right)\right)$$

$$\mathscr{F}^{-1}\left[\text{rect}\left(\frac{k}{K}\right)\right] = \frac{K}{2\pi}\text{sinc}\left(\frac{K}{2\pi}x\right).$$

The complete transform of the window function can thus be written as

$$\mathscr{F}^{-1}[h_K(k)] = \frac{K}{2\pi}\left[\delta(x) + \frac{1}{2}\delta\left(x - \frac{2\pi}{K}\right) + \frac{1}{2}\delta\left(x + \frac{2\pi}{K}\right)\right] * \text{sinc}\left(\frac{K}{2\pi}x\right)$$

$$= \frac{K}{2\pi}\left[\text{sinc}\left(\frac{K}{2\pi}x\right) + \frac{1}{2}\text{sinc}\left(\frac{K}{2\pi}x - 1\right) + \frac{1}{2}\text{sinc}\left(\frac{K}{2\pi}x + 1\right)\right].$$

Compared to the rectangular window, two additional shifted sinc-functions are added to decrease side lobes. To analyze the signal width, the FWHM x_{FWHM} can be computed by

$$\text{sinc}\left(\frac{K}{4\pi}x_{FWHM}\right) + \frac{1}{2}\text{sinc}\left(\frac{K}{4\pi}x_{FWHM} - 1\right) + \frac{1}{2}\text{sinc}\left(\frac{K}{4\pi}x_{FWHM} + 1\right) = \frac{1}{2},$$

which has the trivial solutions

$$x_{FWHM} = \pm\frac{4\pi}{K}. \tag{B.1.1}$$

With a total signal width of K divided in N equal parts of spacing Δk, i.e. $K = N\Delta k$, the resulting spacing in Fourier space can be computed by (2.1.6). It follows for the FWHM

$$x_{FWHM} = N_{FWHM} \cdot \Delta x \approx 4\pi \cdot \frac{1}{K} = 4\pi \cdot \frac{1}{N\Delta k} = 2\Delta x.$$

Thus the number of pixels specifying the FWHM is

$$N_{FWHM} = 2.$$

There are a lot of further windowing functions which have other and/or more shifted sines and cosines to reduce side lobes, similar to the Hann window. However, resolution (FWHM) is always lost.

B.1.3 Ideal Gaussian window and OCT resolution

A Gaussian is an idealized window function. In digital signal processing it cannot be achieved due to its infinite support, i.e. it does not drop to zero for large $|k|$. It is defined by

$$w(k) = g_\sigma(k) = \exp\left(-\frac{k^2}{2\sigma^2}\right),$$

with σ being its $1/e$-intensity-value, i.e. $w^2(\sigma) = 1/e$.

Axial resolution in OCT imaging is most commonly specified for Gaussian spectra in terms of its full-width at half maximum (FWHM) values; for TD-OCT this values is precise, assuming a Gaussian shaped spectrum. The relation to the FWHM value of the Gaussian is $k^2_{FWHM} = 8\ln 2 \cdot \sigma^2$. Its Fourier transform, representing the point spread function and the axial resolution is again a Gaussian

$$\mathscr{F}^{-1}[w(k)] = \frac{1}{\sqrt{2\pi}} \sigma \exp\left(-\frac{x^2}{2(1/\sigma)^2}\right),$$

with its σ-width inverted. Accordingly, the FWHM of this signal is given by

$$x^2_{FWHM} = 8\ln 2 \cdot \frac{1}{\sigma^2} = 64(\ln 2)^2 \frac{1}{k^2_{FWHM}}.$$

The wavenumber k is related to the wavelength by $k = 2\pi/\lambda$. With the differential $dk/d\lambda = -2\pi d\lambda/\lambda^2$ it follows the FWHM formula for a Gaussian PSF

$$x_{FWHM} = \frac{4\ln 2}{\pi} \frac{\lambda_0^2}{\lambda_{FWHM}}.$$

For the resolution in TD-OCT, it additionally needs to be taken into account, that the optical path length is occurring twice, once when illuminating the sample and once after backscattering, the axial resolution is thus

$$x_{FWHM,OCT} = \frac{2\ln 2}{\pi} \frac{\lambda_0^2}{\lambda_{FWHM}}, \qquad (B.1.2)$$

which is valid in time-domain and approximately in FD-OCT for Gaussian shaped spectra.

B.1.4 Truncated Gaussian window

The truncated Gaussian window is an actually feasible version of the Gaussian window in signal processing. It is defined by a certain cut-off frequency K

$$w(k) = \exp\left(-\frac{k^2}{2\sigma^2}\right) \text{rect}\left(\frac{k}{K}\right),$$

where it is forced to zero outside of its slope. Its performance obviously depends on the ratio of σ and the overall bandwidth K. The Fourier transform is computed to

$$\mathscr{F}[w(k)] = \sqrt{2\pi}\sigma \exp\left(-\frac{x^2}{2(1/\sigma)^2}\right) * \operatorname{sinc}\left(\frac{K}{2\pi}x\right)$$

B.1.5 Comparison of window functions

A comparison of window functions is found in the Figures B.1.1 and Figure B.1.2.

B.2 Lateral resolution of an imaging system

B.2.1 Circular aperture

The angular spectrum is limited by an aperture such, that no spatial frequencies $\sqrt{k_x^2 + k_y^2}$ above a certain threshold K contribute to image forming. The numerical aperture (see Section 2.4.5.3) is given by

$$\mathrm{NA} = \frac{K}{k}.$$

The according resolution of the imaging system can be obtained by Fourier transforming the circular limited angular spectrum. In the focus the angular spectrum is supposed to be phase-free, as the Fourier transform of a centered δ-peak is a plane wave.

$$A(k_x, k_y) = \operatorname{circ}\left(\frac{k_x}{K}, \frac{k_y}{K}\right),$$

with circ defined by (A.1.2). The Fourier transform gives the PSF amplitude of the imaging system up to a constant factor and can be computed by (A.3.7) to

$$U(x,y) \propto \mathscr{F}_{xy}^{-1}[A(k_x, k_y)] = \frac{K}{2\pi}\frac{J_1\left(K\sqrt{x^2+y^2}\right)}{\sqrt{x^2+y^2}},$$

with J_1 being the Bessel function of the first kind and first order. It is related to J_0 by

$$J_1(r) = \frac{1}{r}\int_0^r \mathrm{d}r'\, r' J_0(r').$$

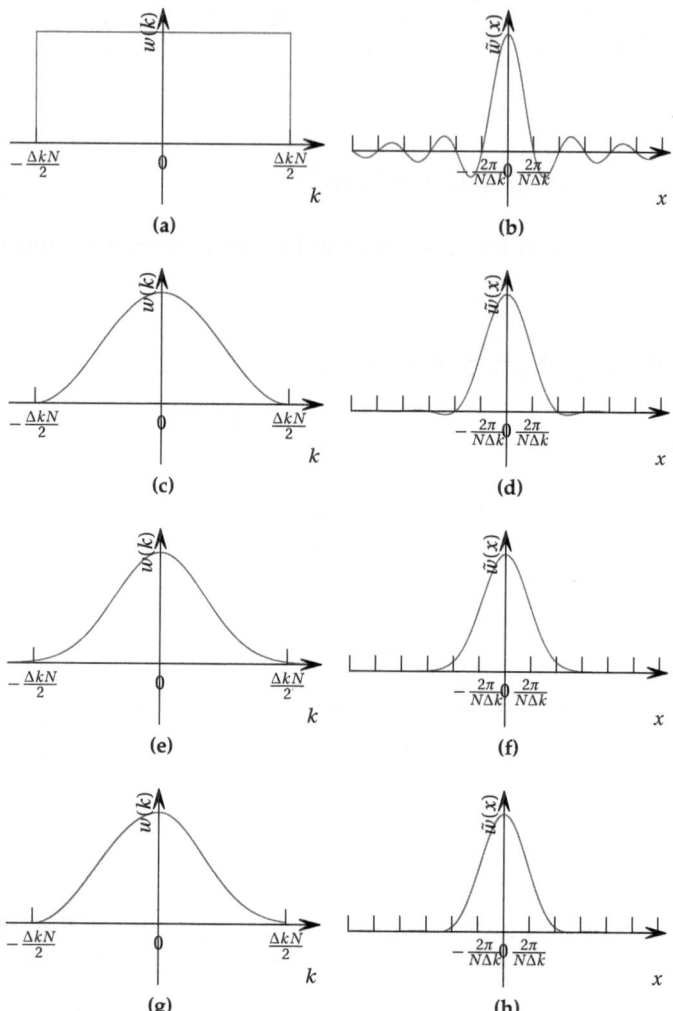

Figure B.1.1.: Axial window functions and their according Fourier transforms. a) Rectangular window. b) Fourier transform of a rectangular window. c) Hann window. d) Fourier transform of a Hann window. e) Ideal Gaussian window function. In practice this window is infinitely extended and cannot be used in digital signal processing. f) Fourier transform of the ideal Gaussian window. g) Truncated Gaussian window. With the truncation, the Gaussian window can be used in digital signal processing. h) Fourier transform of the truncated Gaussian window.

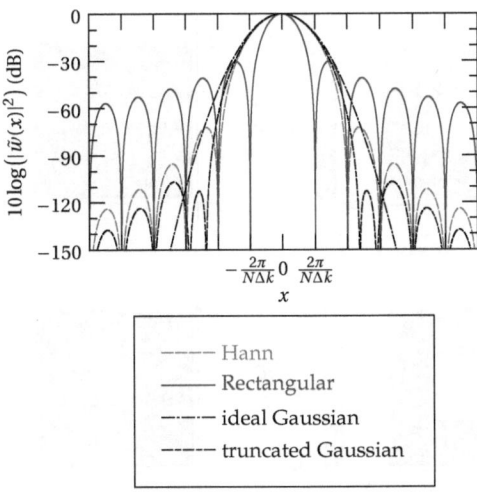

Figure B.1.2.: Axial intensity distribution (PSFs) for various window functions. In general, the smaller the peak width the more severe the side lobes.

The according intensity distribution is in spherical coordinates with radius $r = \sqrt{x^2 + y^2}$ computed to

$$I(r) = \gamma(U^*U)(r) \propto 4\pi^2 K^2 \frac{J_1^2(Kr)}{r^2},$$

known as Airy disk. The resolution is most commonly specified as the distance of the center of the airy disc to its first minimum (Rayleigh criterion, Section 2.4.5.2). It is obtained by computing the first minimum of $J_1(Kr) = 0$ besides $r = 0$, which is the case for $Kr = 3.83$. The resolution is thus

$$r_{res} \approx 3.83 \frac{1}{K} \approx 3.83 \frac{1}{k \cdot NA} \approx 0.61 \frac{\lambda}{NA}. \tag{B.2.1}$$

B.2.2 Rectangular aperture

For a rectangular aperture that is with its borders aligned along the x and y-axis a single NA cannot be specified. If the rectangular aperture filters frequencies with absolute value larger than K_x for the x-axis and K_y for the y-axis, one can define NAs for both axes

$$NA_{x/y} = \frac{K_{x/y}}{k}.$$

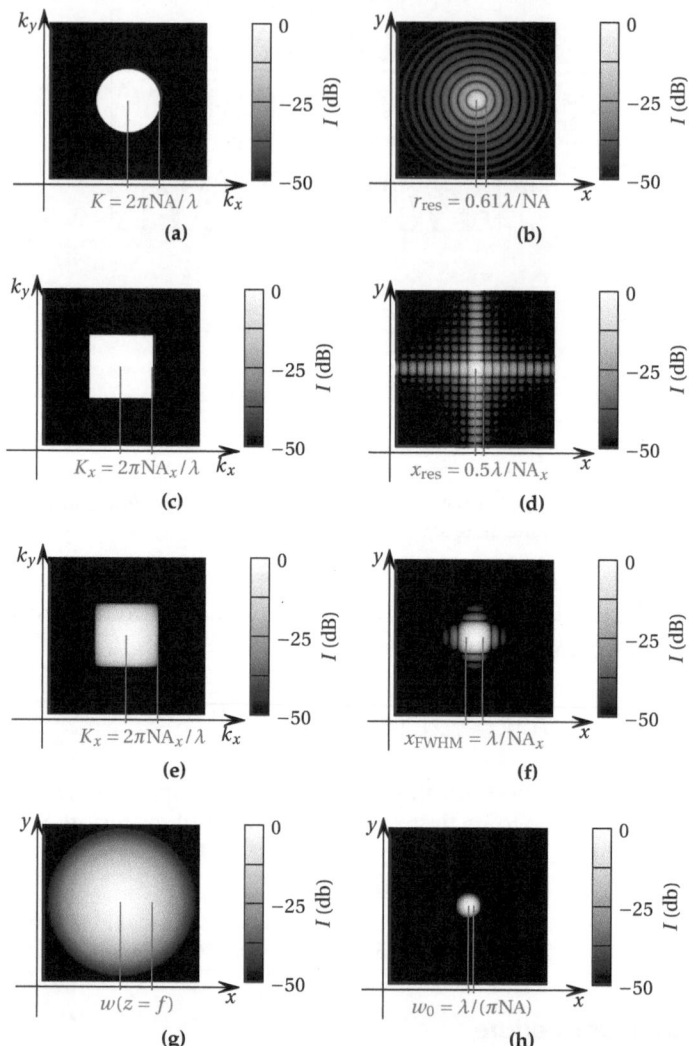

Figure B.2.1.: Apertures in frequency space (a, c, e) or position space (g) and according point spread functions (PSF). a) Circular aperture. b) Airy disk as intensity PSF when using a circular aperture. c) Rectangular aperture. d) The resulting PSF when using a rectangular aperture. Major side lobes are visible. e) Hann apodized beam. f) PSF when using a Hann apodized beam. The lateral resolution is worse but side lobes are decreased. g) Intensity of a Gaussian beam in an out-of-focus layer. h) Intensity of a Gaussian beam in the focus.

The aperture is thus given by

$$A(k_x, k_y) = \text{rect}\left(\frac{k_x}{2K_x}\right) \cdot \text{rect}\left(\frac{k_y}{2K_y}\right),$$

with rect given by (A.1.1). Its Fourier transform can be computed to

$$U(x,y) \propto \mathscr{F}_{xy}\left[A(k_x, k_y)\right] = K_x K_y \, \text{sinc}\left(x\frac{K_x}{\pi}\right) \cdot \text{sinc}\left(y\frac{K_y}{\pi}\right),$$

with sinc given by (A.1.4). The function is not rotationally symmetric. Along the axes of x and y, the imaging system can have a different resolution and in general the resolution will get higher when leaving these axes, reaching its maximum for the diagonal axis of the aperture. We will therefore only have a look at the x-axis. The intensity distribution along the x-axis will be

$$I(x,y) = \gamma(U^*U)(x,y) \propto K_x^2 K_y^2 \text{sinc}^2\left(x\frac{K_x}{\pi}\right) \text{sinc}^2\left(y\frac{K_y}{\pi}\right).$$

Computing the distance of maximum $(x = 0)$ to its first minimum (Rayleigh criterion, Section 2.4.5.2) gives the $\text{sinc}^2(xK/\pi) = 0$ which is maximal for $x = 0$ and minimal for $x = \pi K$. It follows that

$$x_{\text{res}} = \frac{\pi}{K} = \pi\frac{1}{k \cdot \text{NA}_x} = 0.5\frac{\lambda}{\text{NA}_x}. \tag{B.2.2}$$

A respective formula can be derived for y_{res}.

B.2.3 Gaussian beams

The Gaussian function has special unique properties as it is its own Fourier transform. Additionally, it is the function for which the equality sign holds in the uncertainty relation [121] and therefore, it provides optimal resolution for a specific aperture, if both are specified over their standard deviations. Here the field of a Gaussian beams in their paraxial approximation will be derived.

The angular spectrum of a Gaussian field in its focus (2.4.13) is computed to

$$\tilde{U}_{xy}(k_x, k_y, z_0) = A_0 \pi w_0^2 \exp\left(-\frac{w_0^2}{4}\left(k_x^2 + k_y^2\right)\right).$$

Propagating by a distance z gives an additional phase factor, as demonstrated in

Section 2.3.2.2:

$$\tilde{U}_{xy}(k_x, k_y, z_0 + z) = A_0 \pi w_0^2 \exp(ik_z z) \exp\left(-\frac{w_0^2}{4}\left(k_x^2 + k_y^2\right)\right).$$

Using the paraxial approximation of the propagation kernel (2.4.3) this can be rewritten to

$$
\begin{aligned}
\tilde{U}_{xy}(k_x, k_y, z_0 + z) &= A_0 \pi w_0^2 \exp(izk) Q_{-z/k}(k_x, k_y) \exp\left[-\frac{w_0^2}{4}\left(k_x^2 + k_y^2\right)\right] \\
&= A_0 \pi w_0^2 \exp(izk) \exp\left[\left(-\frac{w_0^2}{4} - i\frac{z}{2k}\right)\left(k_x^2 + k_y^2\right)\right].
\end{aligned}
$$

The inverse Fourier transform yields the wave field of a Gaussian beam,

$$U(x, y, z_0 + z) = \frac{A_0 z_R}{-iz - z_R} \exp(izk) \exp\left(-\frac{k}{2}\frac{x^2 + y^2}{-iz - z_R}\right),$$

where the Rayleigh length z_R has been introduced:

$$z_R = \frac{1}{2}kw_0^2 = \frac{\pi w_0^2}{\lambda}. \tag{2.4.14}$$

By writing the first complex factor in its polar representation and introducing the lateral radius $r^2 = x^2 + y^2$ this can be written as

$$U(r, z_0 + z) = A_0 z_R \sqrt{z^2 + z_R^2} \underbrace{\exp\left(-i\arctan\frac{z}{z_R}\right)}_{\text{phase-shift at } z = 0} \underbrace{\exp\left(izk\right)}_{\text{overall phase}}$$

$$\times \underbrace{\exp\left(-i\frac{k}{2}\frac{z}{z^2 + z_R^2}r^2\right)}_{\text{wave-front curvature}} \underbrace{\exp\left(-\frac{k}{2}\frac{z_R}{z^2 + z_R^2}r^2\right)}_{\text{local beam width}}.$$

The first exponential term gives a phase-shift in the focus, known as the Gouy phase shift. It motivates the definition

$$\zeta(z) = \arctan\left(\frac{z}{z_R}\right).$$

The second exponential term gives the overall phase change and is comparable to an infinite plane wave. The third exponential term gives an additional phase, dependent on the lateral position r and thus determines the curvature of the wave-front. It motivates the radius of curvature function

$$R(z) = \frac{z^2 + z_R^2}{z} = z\left(1 + \frac{z_R^2}{z^2}\right).$$

The fourth and last exponential term is a Gaussian and determines the local beam width. It motivates the definition of the local beam width

$$w^2(z) = \frac{2}{k}\frac{z^2 + z_R^2}{z_R} = \frac{2z_R}{k}\left(\frac{z^2}{z_R^2} + 1\right) = w_0^2\left(1 + \frac{z^2}{z_R^2}\right).$$

Using these definitions the wave field of the Gaussian beam can be written as

$$U(r, z_0 + z) = A_0 \frac{w_0}{w(z)} \exp(-i\zeta(z)) \exp\left(izk\right) \exp\left(ik\frac{r^2}{2R(z)}\right) \exp\left(-\frac{r^2}{w^2(z)}\right).$$

In the according intensity distribution $I(r, z)$ the complex parts of this equation cancel. It is given by

$$I(r, z) = \gamma(U^*U)(r, z) = \gamma A_0^2 \frac{w_0^2}{w^2(z)} \exp\left(-\frac{2r^2}{w^2(z)}\right).$$

A schematic drawing of such a Gaussian beam is shown in Figure 2.4.2.

The NA of a Gaussian beam is not well defined as the aperture of the beam does not have sharp boundaries. Most commonly the $1/e^2$ width of the intensity is being used as boundary of the aperture. It is reached if $r = w(z)$ and the NA is given by

$$\mathrm{NA} = \frac{w(z)}{\sqrt{z^2 + w^2(z)}}.$$

Asymptotically, i.e. for large z it follows

$$\frac{z_R^2}{w_0^2} = \frac{1}{\mathrm{NA}^2} - 1.$$

Using the explicit formula for w_0 one obtains formulas for the Rayleigh length and central beam width for a given NA:

$$w_0 = \frac{2}{k}\sqrt{\frac{1}{NA^2} - 1} \approx \frac{2}{kNA} = \frac{\lambda}{\pi NA},$$

where the approximation is only valid for small NA. Additionally, the relation

$$z_R = \frac{2}{k}\left(\frac{1}{NA^2} - 1\right) \approx \frac{2}{kNA^2} = \frac{\lambda}{\pi NA^2} = \frac{w_0}{NA},$$

where again, the approximation is only valid for small NA. However, for large NA the paraxial approximation itself does not hold.

C

Discretization of the holoscopy reconstruction process

In the following Appendix the discretization of the holoscopy reconstruction process is outlined briefly. It is required for actual implementation of the algorithms. A few additional considerations need to be applied that are not obvious from the analytical reconstruction process and are considered here.

C.1 General considerations

C.1.1 Sampling theorem for discretization of analytical functions

According to the sampling theorem, a bandlimited signal with bandlimit K needs to have at least a sampling frequency of $2K$ in order to be sampled correctly (see Section 2.2.2). While this identity is important when actually sampling physical signals, it also needs to be taken into account during numerical computations, especially if analytical functions are discretized. As an example, consider a spherical wave given by

$$U(x,y) = \exp\left(ik\sqrt{x^2 + y^2 + z_0^2} \right) \qquad (C.1.1)$$

to be used numerically. Local frequencies are most easily obtained by looking at the Taylor expansion of the argument around central x_0 and y_0 [49]:

$$\arg U(x,y) \approx \arg U(x_0,y_0) + \left. \frac{\partial}{\partial x} \arg U(x,y) \right|_{x_0,y_0} (x - x_0)$$

$$+ \left. \frac{\partial}{\partial y} \arg U(x,y) \right|_{x_0,y_0} (y - y_0)$$

In case of the spherical wave (C.1.1) this is

$$k\sqrt{x^2 + y^2 + z_0^2}$$

$$\approx k\sqrt{x_0^2 + y_0^2 + z_0^2} + k\frac{1}{\sqrt{x_0^2 + y_0^2 + z_0^2}}(x - x_0) + k\frac{1}{\sqrt{x_0^2 + y_0^2 + z_0^2}}(y - y_0),$$

i.e. the (absolute) occurring frequencies are

$$k_x = k_y = \frac{k}{\sqrt{x_0^2 + y_0^2 + z_0^2}}.$$

The discretized version of the wave field (C.1.1) is

$$u_\ell = \exp\left(ik\sqrt{\left(\ell - \frac{L_0}{2} \right)^2 \Delta x^2 + \left(m - \frac{M_0}{2} \right)^2 \Delta y^2 + z_0^2} \right),$$

with pixel spacings Δx and Δy and pixel index $\ell = 0, ..., L_0 - 1$ and $m = 0, ..., M_0 - 1$. The maximum local frequencies are obtained in the corners of the image, because of the symmetry it suffices to look at one corner for one of two frequencies, for example k_x at $\ell = 0$ and $m = 0$:

$$k_{x,\max} = \frac{k}{\sqrt{(L_0^2 \Delta x^2 + M_0^2 \Delta y^2)/4 + z_0^2}}.$$

C.1.2 Bandwidth of a convolution

Having two signals with frequencies k_1 and k_2, for example $\exp(ik_1x)$ and $\exp(ik_2x)$, their product $\exp(i(k_1+k_2)x)$ has frequency $k_1 + k_2$. Therefore, the product of two function $f(x)$ and $g(x)$ with bandlimits K_f and K_g, i.e.

$$\begin{aligned} \tilde{f}(k) &= 0, \quad \text{for } |k| > K_f \\ \tilde{g}(k) &= 0, \quad \text{for } |k| > K_g \end{aligned}$$

has a bandlimit $K_f + K_g$. In fact, the product can be written as

$$f(x) \cdot g(x) = \frac{1}{(2\pi)^2} \int dk \int dk' \, e^{i(k+k')x} \tilde{f}(k)\tilde{g}(k'),$$

and the frequencies in this multiplication are $k + k'$ which is for bandlimited f and g limited by $K_f + K_g$. Consequently, the opposite is also true: the convolution of two spacelimited functions has added spacelimits, which results in a circular convolution. In case of discretized finite signals, adding and convolving of space or bandlimited signals have profound consequences: aliasing artifacts can occur, whenever discrete representations of signals and functions are multiplicated or convolved. This aliasing can occur in frequency or position space. In the reconstruction of holoscopic signals, the multiplication with lenses, reference waves and the propagation, which is essentially a convolution, therefore need to be treated carefully.

C.1.3 Zero padding

To prevent aliasing artifacts zero-padding can be applied. Zero-padding in the frequency domain increases the Nyquist frequency in position space and vice versa, such that convolutions and multiplications are safe. For example, the angular spectrum approach is equivalent to a convolution (Section 2.3.2.2) and due to its effective implementation using fast Fourier transforms the convolution carried out is circular – or, from an equivalent point of view, aliasing artifacts can occur. Effectively, this shows up as light rays that reach the image border are wrapped around and appear on the other side of the image, instead of being propagated outwards (see e.g. [53]). The simplest solution to prevent the circular convolution is zero-padding of the input signals, i.e. using a larger image field array with empty space outwards of the area of interest. This way the light rays have space to propagate to and do not disturb the original image. Other techniques for computing the convolution – such as the overlap-add or the overlap-save methods (see e.g. [122]) – can also be applied, but are in general significantly more complicated to implement.

C.2 The acquired image and wave fields

C.2.1 Obtaining the object wave-field

The digital images $i_{\ell m n'}$, as acquired by the camera, are related to the analytical images $I(x, y, t(k))$ by

$$i_{\ell m n'} = I\left(\left(\ell - \frac{L_0}{2}\right) \cdot \Delta x_0, \left(m - \frac{M_0}{2}\right) \cdot \Delta y_0, k_0 + n'\Delta k\right),$$

where the spatial coordinates are $x = \left(\ell - \frac{L_0}{2}\right)\Delta x_0$, $y = \left(m - \frac{M_0}{2}\right)\Delta y_0$ and $k = k_0 + n'\Delta k$. The range of the indices ℓ and m depends on the number of pixels being used and the range of n' depends on the number of images acquired during the sweep. We assume a total of N images with $L_0 \times M_0$ pixels each, and hence $\ell = 0, ..., L_0 - 1$, $m = 0, ..., M_0 - 1$ and $n' = 0, ..., N - 1$. Δx_0 and Δy_0 are given by the pixel size of the camera and Δk is given by the sweep range (see Section 5.3.1) according to

$$\Delta k = \frac{k_f - k_i}{N - 1},$$

with k_i and k_f being the initial and final wavenumber of the sweep, respectively.

A first suppression of the DC signals and autocorrelated terms can be done according to Section 5.3.3.1 by using the averaged images and spectra given by

$$i_{\overline{\ell m} n'} = \frac{1}{L_0 M_0} \sum_{\ell=0}^{L_0-1} \sum_{m=0}^{M_0-1} i_{\ell m n'}$$

$$i_{\ell m \overline{n'}} = \frac{1}{N} \sum_{n=0}^{N-1} i_{\ell m n'}$$

$$i_{\overline{\ell m n'}} = \frac{1}{L_0 M_0 N} \sum_{\ell=0}^{L_0-1} \sum_{m=0}^{M_0-1} \sum_{n=0}^{N-1} i_{\ell m n'}.$$

The DC-corrected image field is then given by

$$i_{\text{corr},\ell m n'} = w_{\ell m n'} \frac{i_{\ell m n'} - i_{\overline{\ell m} n'} - i_{\ell m \overline{n'}} + i_{\overline{\ell m n'}}}{i_{\overline{\ell m} n'} i_{\ell m \overline{n'}}},$$

with $w_{\ell mn'}$ being a suitable window function. In numerical computations it needs to be assured that $i_{\overline{\ell mn'}} \neq 0$ and $i_{\ell \overline{mn'}} \neq 0$ for all ℓ, m and n'.

C.2.2 Time-frequency filter

In an on-axis geometry, removing the negative frequency components can most easily be done by applying the filter in time-frequency (Section 5.3.3.2), i.e. wavenumber space by

$$i_{\ell mn'}^{\mathcal{H}} = 2 \sum_{n=0}^{N-1} F_{n'n}^{-1} \Theta_n \left(\sum_{n'=0}^{N-1} F_{n'n} i_{\text{corr},\ell mn'} \right)$$

with

$$\Theta_n = \begin{cases} 0 & \text{for } n > \frac{N}{2} \\ 1 & \text{otherwise} \end{cases}$$

and the Fourier matrix and its inverse being given according to Section 2.1.2 by

$$F_{n'n} = \exp\left(-i\frac{2\pi}{N} n'n \right) \quad \text{and} \quad F_{n'n}^{-1} = \frac{1}{N} \exp\left(i\frac{2\pi}{N} n'n \right).$$

The index n' is used twice here, once summed over, to ease readability. The same will be used in the following Sections when applicable. For the time-frequency filter the pixel spacing and dimensions of the volume in general remain the same, i.e.

$$L = L_0 \quad \text{and} \quad M = M_0$$

and

$$\Delta x = \Delta x_0 \quad \text{and} \quad \Delta y = \Delta y_0,$$

with $L \times M$ being the size of images and Δx and Δy being the spacing of the pixels after the filter .

C.2.3 Removing the reference wave field

The discrete form of the phase-corrected spherical reference wave field (5.3.6) is given by

$$r_{\ell mn'} = R_0 \left(\left(\ell - \frac{L}{2} \right) \cdot \Delta x, \left(m - \frac{M}{2} \right) \cdot \Delta y, k_0 + n'\Delta k \right)$$

$$\propto \exp\left[-\mathrm{i}(k_0 + n'\Delta k) \right.$$

$$\left. \times \left(\sqrt{\left(\ell - \frac{L}{2}\right)^2 \Delta x^2 + \left(m - \frac{M}{2}\right)^2 \Delta y^2 + (f + z_0)^2} - f - z_0 \right) \right].$$

The discretized phase-corrected object field $o_{lmn'}$ and its analytical representation $O_0(x, y, k)$ (Section 5.3.2) are connected by

$$o_{\ell m n'} \;=\; O_0\!\left(\left(\ell - \frac{L}{2}\right)\cdot \Delta x, \left(m - \frac{M}{2}\right)\cdot \Delta y, k_0 + n'\cdot \Delta k \right).$$

The former can be obtained by the discretization of (5.3.12)

$$o_{\ell m n'} = \frac{i\mathcal{H}_{\ell m n'}^{\ell}}{r_{\ell m n'}},$$

where the proportionality factor γ has been ignored.

C.2.4 Spatial filter

As in digital holography, spatial filters are used to optimize imaging in holoscopy. If not all spatial frequencies of the acquired image data contain meaningful fringes, but noise, reflections or DC parts of the acquired signals, the resulting artifacts can often be effectively removed using suitable spatial filters. Especially, in off-axis geometries, the DC-part, the autocorrelation term of the object wave field, and fixed pattern noise can be filtered (Section 5.3.3.2).

For an off-axis geometry a filter can be supplied in spatial frequency domain reducing the DC part of the image and autocorrelation terms of the object. In general afterwards only a subregion of the original image is used, reducing the image size from $L_0 \times M_0$ to $L \times M$.

The filter is achieved by

$$i_{\ell m n'}^{\text{filtered}} = \sum_{\ell''=0}^{L-1}\sum_{m''=0}^{M-1} F_{\ell''\ell m''m}^{-1}\,\Theta_{\ell''m''}\left[\sum_{\ell'=0}^{L_0-1}\sum_{m'=0}^{M_0-1} S_{\ell''\ell'm''m'}\left(\sum_{\ell=0}^{L_0-1}\sum_{m=0}^{M_0-1} F_{\ell'\ell m'm}\,i_{\ell m n'} \right) \right]$$

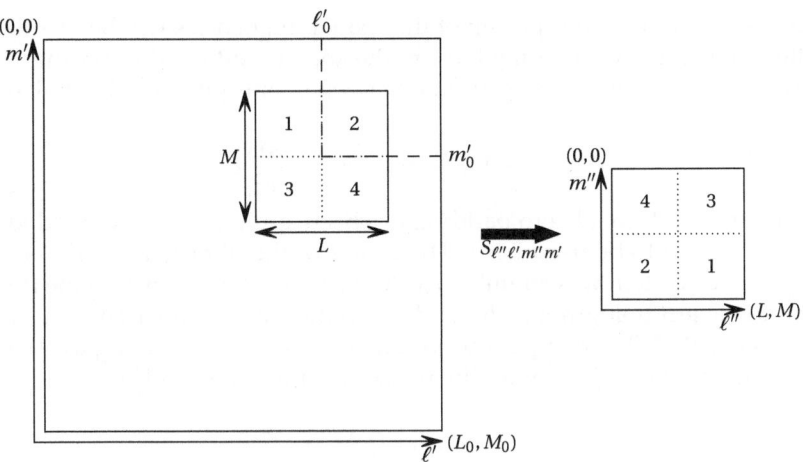

Figure C.2.1.: Filtering and resorting of data points with the tensor $S_{\ell''\ell m''m}$. Using the common definition of the discrete Fourier transform (2.1.2), the frequency $(k_x, k_y) = (0,0)$ is found in pixel coordinate $(\ell', m') = (0,0)$ in frequency space. In the windowed region the frequency $(k_x, k_y) = (0,0)$ is centered and thus resorting of the data quadrants is required prior to the inverse Fourier transform.

with

$$S_{\ell''\ell'm''m'} = \begin{cases} \delta_{\ell'',\ell'-\ell'_0}\delta_{m'',m'-m'_0} & \text{for } 0 \le \ell'' < \frac{L}{2} \text{ and } 0 \le m'' < \frac{M}{2} \\ \delta_{\ell'',\ell'-\ell'_0}\delta_{m''-M,m'-m'_0} & \text{for } 0 \le \ell'' < \frac{L}{2} \text{ and } \frac{M}{2} \le m'' < M \\ \delta_{\ell''-L,\ell'-\ell'_0}\delta_{m'',m'-m'_0} & \text{for } \frac{L}{2} \le \ell'' < L \text{ and } 0 \le m'' < \frac{M}{2} \\ \delta_{\ell''-L,\ell'-\ell'_0}\delta_{m''-M,m'-m'_0} & \text{for } \frac{L}{2} \le \ell'' < L \text{ and } \frac{M}{2} \le m'' < M \\ 0 & \text{otherwise,} \end{cases}$$

and $\Theta_{\ell''m''}$ being a suitable window function in the range $0 \le \ell'' < L$ and $0 \le m'' < M$ and ℓ'_0, m'_0 giving the center of the filter. $F_{\ell'\ell m'm}$ denotes the Fourier tensor for a field of size $L_0 \times M_0$ and $F^{-1}_{\ell''\ell m''m}$ the inverse Fourier tensor for a size $L \times M$ (compare Section 2.1.2, especially (2.1.3)). It should be noted that in this formula the indices ℓ and m have been used twice, once as spatial coordinate that is summed over and therefore does not influence the result. As this is more readable and there is no danger of confusion, it will be used this way in the next Sections as well. The effect of the filtering operation and of the tensor $S_{\ell''\ell m''m}$ is illustrated in Figure C.2.1.

In frequency domain the spacing of the spatial frequencies will be identical after this filter. However, when going back to the spatial domain the spacing of pixels will have increased due to the reduced image size. Thus pixel spacing will be given by

$$\Delta x = \frac{L_0}{L}\Delta x_0 \quad \text{and} \quad \Delta y = \frac{M_0}{M}\Delta y_0.$$

Assuming that additional zero padding has been applied, the zero padded values for L and M need to be used here. The center of the filter ℓ'_0 and m'_0 are usually determined at a certain wavenumber k_0. According to Section 5.3.3.3 the frequency offset of the object hologram is dependent on the wavenumber and thus it needs to be compensated. The compensation can be done by multiplying the intensities in the spatial domain $i^{\text{filtered}}_{\ell mn'}$ with the phase function (see (5.3.15))

$$\Phi_{\ell m} = e^{-i2\pi\left(\mp 1 - \frac{k}{k_0}\right)\left(\frac{\ell\ell'_0}{L} + \frac{mm'_0}{M}\right)}.$$

The resulting field is then given by

$$i^{\text{filtered}}_{\ell mn'} = \Phi_{\ell m} i^{\text{filtered}}_{\ell mn'}.$$

The effect of the phase function $\Phi_{\ell m}$ is illustrated in Figure 5.3.4 for two holograms acquired at different wavelengths.

C.3 The angular spectrum

The plane wave expansion of $o_{\ell mn'}$ denoted $\tilde{o}_{\ell'm'n'}$ for all acquired holograms $n' = 0, \ldots, N-1$ can be obtained by a two-dimensional Fourier transform that can be computed using the two-dimensional Fourier tensor (2.1.3), i.e. by

$$\tilde{o}_{\ell'm'n'} = \sum_{\ell=0}^{L-1}\sum_{m=0}^{M-1} F_{\ell'\ell m'm}\, o_{\ell mn'}.$$

The pixel spacing in the angular spectrum is given by

$$\Delta k_x = \frac{2\pi}{L\Delta x} \quad \text{and} \quad \Delta k_y = \frac{2\pi}{M\Delta y}.$$

As generalization of (2.1.4), the symmetries of the Fourier tensor are given by

$$F_{\ell'\ell m'm} = F_{(\ell'+L)\ell m'm} = F_{\ell'(\ell+L)m'm} = F_{\ell'\ell(m'+M)m} = F_{\ell'\ell m'(m+M)}. \qquad (\text{C.3.1})$$

Going back to the original space can be achieved by using the inverse Fourier tensor

$$F^{-1}_{\ell'\ell m'm} = \frac{1}{LM} \exp\left[+i\left(\frac{2\pi}{L}\ell\ell' + \frac{2\pi}{M}mm'\right)\right].$$

C.3.1 Propagation

Phase-corrected propagation (see Section 5.3.2 and 2.3.2.2) of the angular spectrum of a wave field is performed by multiplying it with the propagation function $P^0_{\ell'm'}(k,z)$ that is given by

$$P^0_{\ell'm'}(k,z) = \exp\left[iz\sqrt{k^2 - (\ell' \cdot \Delta k_x)^2 - (m' \cdot \Delta k_y)^2} - izk\right]$$
$$\text{for } \ell = -\tfrac{L}{2}, ..., \tfrac{L}{2} - 1, m = -\tfrac{M}{2}, ..., \tfrac{M}{2} - 1.$$

The symmetry of the Fourier tensor (C.3.1) causes a symmetry in the indices of the propagation function given by

$$P^0_{\ell'm'}(k,z) = P^0_{(\ell'-L)m'}(k,z) = P^0_{\ell'(m'-M)}.$$

This symmetry can be used to compute the elements for $\ell = \tfrac{L}{2}, ..., L-1$ and $m = \tfrac{M}{2}, ..., M-1$.

In spatial coordinates the propagation is described by

$$\mathscr{P}_{k,z}[0_{\ell m n'}] = \sum_{\ell'=0}^{L-1}\sum_{m'=0}^{M-1} F^{-1}_{\ell'\ell m'm} P^0_{\ell'm'}(k,z) \left(\sum_{\ell=0}^{L-1}\sum_{m=0}^{M-1} F_{\ell'\ell m'm} 0_{\ell m n'}\right).$$

C.4 Reconstruction

C.4.1 Single reconstruction

The single reconstruction is performed by applying a one-dimensional Fourier transform on the propagated wave field, i.e.

$$\int dk\, e^{-i2kz}\, \mathscr{P}^0_{k,-z_0-z_p}[\gamma O_0(x,y,t(k))].$$

In discrete form this can be written as

$$\eta_{\ell mn} = \sum_{n'=0}^{N-1} F_{n'n} \mathscr{P}^0_{k,-z_0-z_p} [o_{\ell mn'}],$$

with $\eta_{\ell mn}$ being the discrete scattering potential. The Fourier matrix here is given by

$$
\begin{aligned}
F_{n'n} &= \exp[i2kz] \\
&= \exp\left[i\left(2\left(k_i + n'\Delta k\right)\cdot n\Delta z\right)\right] \\
&= \exp[i2k_i\Delta zn]\exp\left[i\left(\frac{2\pi}{N}\left(2\frac{N}{2\pi}\Delta k\Delta z\right)n'\cdot n\right)\right] \\
&= \exp[i2k_i\Delta zn]\exp\left[i\frac{2\pi}{N}n'n\right].
\end{aligned}
$$

In this Fourier matrix wavenumber spacing Δk and spatial coordinate spacing Δz depend on each other by

$$\Delta z = \frac{\pi}{N\Delta k},$$

as in FD-OCT.

C.4.2 Complete one-step reconstruction

To achieve a complete reconstruction, the object wave fields need to be propagated to the reference plane first. This is achieved by

$$f_{\ell'm'n'} = P^0_{\ell'm'}(k,-z_0)\tilde{o}_{\ell'm'n'}.$$

Afterwards, the integral (5.4.14) has to be performed and yields the angular spectrum of the scattering potential $\tilde{\eta}$. After discretization it can be rewritten as

$$\tilde{\eta}_{\ell'm'n} = \sum_{n'=0}^{N-1} D_{\ell'm'n'n} f_{\ell'm'n'}, \quad \text{with } D_{\ell'm'n'n} = \exp\left(-i\frac{2\pi}{N}v'_{\ell'm'}(n')n\right), \quad \text{(C.4.1)}$$

where $f_{\ell'm'n'}$ is the discretized function $f(k_x,k_y,k)$, and $v'_{\ell'm'}(n')$ can be computed to

$$v'_{\ell'm'}(n') = N\frac{\kappa_\zeta(\ell'\Delta k_x, m'\Delta k_y; k_0 + n'\Delta k) - \kappa_{\zeta,\min}}{\kappa_{\zeta,\max} - \kappa_{\zeta,\min}}.$$

Here, $\kappa_\zeta(k_x, k_y, k)$, $\kappa_{\zeta,\min}$, and $\kappa_{\zeta,\max}$ are given by (5.4.10), (5.4.12), and (5.4.13), respectively. Effective implementations of the NDFT described by (C.4.1) were evaluated in Section 2.1.3 and compared in Chapter 3. These algorithms provide a combined resampling in Fourier space and fast Fourier transform. Finally, the reconstruction is then obtained by two-dimensional Fourier transforming the result back to its original space, i.e.

$$\eta_{\ell m n} = \sum_{\ell'=0}^{L-1} \sum_{m'=0}^{M-1} F_{\ell'\ell m' m}^{-1} \tilde{\eta}_{\ell' m' n}.$$

For the resulting data, the axial spacing of the recreated layers will be given by

$$\Delta z = \frac{2\pi}{N \Delta \kappa_\zeta}, \tag{C.4.2}$$

with

$$\Delta \kappa_\zeta = \frac{\kappa_{\zeta,\max} - \kappa_{\zeta,\min}}{N}.$$

Bibliography

[1] G. N. Hounsfield, *Computed Medical Imaging: Nobel Lecture in physiology or medicine 1971-1980* (1979).

[2] H. Jörnvall, *Physiology Or Medicine, 2001-2005*, Nobel Lectures in Physiology or Medicine (World Scientific, 2008).

[3] D. Huang, E. A. Swanson, C. P. Lin, J. S. Schuman, W. G. Stinson, W. Chang, M. R. Hee, T. Flotte, K. Gregory, C. A. Puliafito, and J. G. Fujimoto, "Optical coherence tomography," Science **254**, 1178–1181 (1991).

[4] E. A. Swanson, J. A. Izatt, M. R. Hee, D. Huang, C. P. Lin, J. S. Schuman, C. A. Puliafito, and J. G. Fujimoto, "In vivo retinal imaging by optical coherence tomography," Opt. Lett. **18**, 1864–1866 (1993).

[5] A. F. Fercher, C. K. Hitzenberger, W. Drexler, G. Kamp, and H. Sattmann, "In vivo optical coherence tomography," Am. J. Ophthalmol. **116**, 113–114 (1993).

[6] M. Wojtkowski, R. Leitgeb, A. Kowalczyk, T. Bajraszewski, and A. F. Fercher, "In vivo human retinal imaging by fourier domain optical coherence tomography," J. Biomed. Opt. **7**, 457–463 (2002).

[7] M. J. Maldonado, L. Ruiz-Oblitas, J. M. Munuera, D. Aliseda, A. Garcia-Layana, and J. Moreno-Montanes, "Optical coherence tomography evaluation of the corneal cap and stromal bed features after laser in situ keratomileusis for high myopia and astigmatism," Ophthalmology **107**, 81–87 (2000).

[8] J. Jungwirth, B. Baumann, M. Pircher, E. Götzinger, and C. K. Hitzenberger, "Extended in vivo anterior eye-segment imaging with full-range complex spectral domain optical coherence tomography," J. Biomed. Opt. **14**, 050501–050503 (2009).

[9] G. Häusler and M. W. Lindner, ""Coherence Radar" and "Spectral Radar"—new tools for dermatological diagnosis," J. Biomed. Opt. **3**, 21–31 (1998).

[10] C. Blatter, J. Weingast, A. Alex, B. Grajciar, W. Wieser, W. Drexler, R. Huber, and R. A. Leitgeb, "In situ structural and microangiographic assessment of human skin lesions with high-speed oct," Biomed. Opt. Express **3**, 2636–2646 (2012).

[11] B. J. Vakoc, D. Fukumura, R. K. Jain, and B. E. Bouma, "Cancer imaging by optical coherence tomography: preclinical progress and clinical potential," Nat. Rev. Cancer **12**, 363–368 (2012).

[12] E. Lankenau, D. Klinger, C. Winter, A. Malik, H. Müller, S. Oelkers, H.-W. Pau, T. Just, and G. Hüttmann, "Combining optical coherence tomography (OCT) with an operating microscope," in *Advances in Medical Engineering*, T. M. Buzug, D. Holz, S. Weber, J. Bongartz, M. Kohl-Bareise, and U. Hartmann, eds. (Springer, Berlin, Heidelberg, New York, 2007), pp. 343–348.

[13] G. Hüttmann, J. Probst, T. Just, H. Pau, S. Oelckers, D. Hillmann, P. Koch, and E. Lankenau, "Real-time volumetric optical coherence tomography OCT imaging with a surgical microscope," Head Neck Oncol. **2**, O8 (2010).

[14] J. P. Ehlers, Y. K. Tao, S. Farsiu, R. Maldonado, J. A. Izatt, and C. A. Toth, "Integration of a spectral domain optical coherence tomography system into a surgical microscope for intraoperative imaging," Invest. Ophthalmol. Vis. Sci. **52**, 3153–3159 (2011).

[15] J. Zhang, Z. Chen, and G. Isenberg, "Gastrointestinal optical coherence tomography: Clinical applications, limitations, and research priorities," Gastrointest. Endosc. Clin. N. Am. **19**, 243–259 (2009).

[16] J. W. Villard, K. K. Cheruku, and M. D. Feldman, "Applications of optical coherence tomography in cardiovascular medicine, Part 1," J. Nucl. Cardiol. **16**, 287–303 (2009).

[17] J. W. Villard, A. S. Paranjape, D. A. Victor, and M. D. Feldman, "Applications of optical coherence tomography in cardiovascular medicine, Part 2," J. Nucl. Cardiol. **16**, 620–639 (2009).

[18] P. Antoniuk, M. Strąkowski, J. Pluciński, and B. Kosmowski, "Non-destructive inspection of anti-corrosion protective coatings using optical coherent tomography," Metrol. Meas. Syst. **19**, 365–372 (2012).

[19] K. Wiesauer, M. Pircher, E. Götzinger, C. Hitzenberger, R. Oster, and D. Stifter, "Investigation of glass-fibre reinforced polymers by polarisation-sensitive, ultra-high resolution optical coherence tomography: Internal structures, defects and stress," Compos. Sci. Technol. **67**, 3051–3058 (2007).

[20] D. Stifter, K. Wiesauer, M. Wurm, E. Schlotthauer, J. Kastner, M. Pircher, E. Götzinger, and C. Hitzenberger, "Investigation of polymer and polymer/fibre composite materials with optical coherence tomography," Meas. Sci. Technol. **19**, 074011 (2008).

[21] A. Nemeth, R. Gahleitner, G. Hannesschläger, G. Pfandler, and M. Leitner, "Ambiguity-free spectral-domain optical coherence tomography for determining the layer thicknesses in fluttering foils in real time," Opt. Laser Eng. **50**, 1372–1376 (2012).

[22] L. Thrane, T. M. Jørgensen, M. Jørgensen, and F. C. Krebs, "Application of optical coherence tomography (OCT) as a 3-dimensional imaging technique for roll-to-roll coated polymer solar cells," Sol. Energy Mater. Sol. Cells **97**, 181–185 (2012).

[23] A. Dubois, L. Vabre, A.-C. Boccara, and E. Beaurepaire, "High-resolution full-field optical coherence tomography with a Linnik microscope," Appl. Opt. **41**, 805–812 (2002).

[24] A. Dubois, K. Grieve, G. Moneron, R. Lecaque, L. Vabre, and C. Boccara, "Ultrahigh-resolution full-field optical coherence tomography," Appl. Opt. **43**, 2874–2883 (2004).

[25] K. Grieve, A. Dubois, M. Simonutti, M. Paques, J. Sahel, J.-F. L. Gargasson, and C. Boccara, "In vivo anterior segment imaging in the rat eye with high speed white light full-field optical coherence tomography," Opt. Express **13**, 6286–6295 (2005).

[26] J. Binding, J. B. Arous, J.-F. Léger, S. Gigan, C. Boccara, and L. Bourdieu, "Brain refractive index measured in vivo with high-na defocus-corrected full-field oct and consequences for two-photon microscopy," Opt. Express **19**, 4833–4847 (2011).

[27] J. Holmes, "Theory and applications of multi-beam OCT," Proc. SPIE **7139**, 713908–713907 (2008).

[28] R. A. Leitgeb, M. Villiger, A. H. Bachmann, L. Steinmann, and T. Lasser, "Extended focus depth for Fourier domain optical coherence microscopy," Opt. Lett. **31**, 2450–2452 (2006).

[29] L. Liu, C. Liu, W. C. Howe, C. J. R. Sheppard, and N. Chen, "Binary-phase spatial filter for real-time swept-source optical coherence microscopy," Opt. Lett. **32**, 2375–2377 (2007).

[30] K.-S. Lee and J. P. Rolland, "Bessel beam spectral-domain high-resolution optical coherence tomography with micro-optic axicon providing extended focusing range," Opt. Lett. **33**, 1696–1698 (2008).

[31] C. Blatter, B. Grajciar, C. M. Eigenwillig, W. Wieser, B. R. Biedermann, R. Huber, and R. A. Leitgeb, "Extended focus high-speed swept source OCT with self-reconstructive illumination," Opt. Express **19**, 12141–12155 (2011).

[32] C. Blatter, B. Grajciar, C. M. Eigenwillig, W. Wieser, B. R. Biedermann, R. Huber, and R. A. Leitgeb, "High-speed functional OCT with self-reconstructive Bessel illumination at 1300 nm," Proc. SPIE **8091**, 809104 (2011).

[33] T. S. Ralston, D. L. Marks, P. S. Carney, and S. A. Boppart, "Interferometric synthetic aperture microscopy: Inverse scattering for optical coherence tomography," Opt. Photon. News **17**, 25–25 (2006).

[34] T. S. Ralston, D. L. Marks, P. Scott Carney, and S. A. Boppart, "Interferometric synthetic aperture microscopy," Nat. Phys. **3**, 129–134 (2007). 10.1038/nphys514.

[35] T. S. Ralston, D. L. Marks, P. S. Carney, and S. A. Boppart, "Real-time interferometric synthetic aperture microscopy," Opt. Express **16**, 2555–2569 (2008).

[36] L. Yu, B. Rao, J. Zhang, J. Su, Q. Wang, S. Guo, and Z. Chen, "Improved lateral resolution in optical coherence tomography by digital focusing using two-dimensional numerical diffraction method," Opt. Express **15**, 7634–7641 (2007).

[37] A. A. Moiseev, G. V. Gelikonov, P. A. Shilyagin, D. A. Terpelov, and V. M. Gelikonov, "Digital refocusing in optical coherence tomography," Proc. SPIE **8213**, 82132C (2012).

[38] B. Považay, A. Unterhuber, B. Hermann, H. Sattmann, H. Arthaber, and W. Drexler, "Full-field time-encoded frequency-domain optical coherence tomography," Opt. Express **14**, 7661–7669 (2006).

[39] T. Bonin, G. Franke, M. Hagen-Eggert, P. Koch, and G. Hüttmann, "In vivo Fourier-domain full-field OCT of the human retina with 1.5 million A-lines/s," Opt. Lett. **35**, 3432–3434 (2010).

[40] A. V. Zvyagin, "Fourier-domain optical coherence tomography: optimization of signal-to-noise ratio in full space," Opt. Commun. **242**, 97–108 (2004).

[41] A. V. Zvyagin, P. Blazkiewicz, and J. Vintrou, "Image reconstruction in full-field Fourier-domain optical coherence tomography," J. Opt. A - Pure Appl. Op. **7**, 350 (2005).

[42] D. V. Shabanov, G. V. Geliknov, and V. M. Gelikonov, "Broadband digital holographic technique of optical coherence tomography for 3-dimensional biotissue visualization," Laser Phys. Lett. **6**, 753–758 (2009).

[43] M. K. Kim, "Wavelength-scanning digital interference holography for optical section imaging," Opt. Lett. **24**, 1693–1695 (1999).

[44] F. Montfort, T. Colomb, F. Charrière, J. Kühn, P. Marquet, E. Cuche, S. Herminjard, and C. Depeursinge, "Submicrometer optical tomography by multiple-wavelength digital holographic microscopy," Appl. Opt. **45**, 8209–8217 (2006).

[45] M. C. Potcoava and M. K. Kim, "Optical tomography for biomedical applications by digital interference holography," Meas. Sci. Technol. **19**, 074010 (2008).

[46] D. L. Marks, T. S. Ralston, S. A. Boppart, and P. S. Carney, "Inverse scattering for frequency-scanned full-field optical coherence tomography," J. Opt. Soc. Am. A **24**, 1034–1041 (2007).

[47] B. J. Davis, D. L. Marks, T. S. Ralston, P. S. Carney, and S. A. Boppart, "Interferometric synthetic aperture microscopy: Computed imaging for scanned coherent microscopy," Sensors **8**, 3903–3931 (2008).

[48] R. Gray and J. Goodman, *Fourier Transforms: An Introduction for Engineers*, The Kluwer international series in engineering and computer science (Kluwer Academic Publishers, 1996).

[49] J. Goodman, *Introduction to Fourier optics*, McGraw-Hill physical and quantum electronics series (Roberts & Co., 2005).

[50] W. Lauterborn and T. Kurz, *Coherent Optics: Fundamentals and Applications* (Springer-Verlag, 1998).

[51] M. Born, E. Wolf, A. Bhatia, P. Clemmow, D. Gábor, A. Stokes, A. Taylor, P. Wayman, and W. Wilcock, *Principles of Optics: Electromagnetic Theory of Propagation, Interference and Diffraction of Light* (Cambridge University Press, 2000).

[52] U. Schnars and W. Jueptner, *Digital holography: digital hologram recording, numerical reconstruction, and related techniques* (Springer, 2005).

[53] M. Kim, *Digital Holographic Microscopy: Principles, Techniques, and Applications*, Springer Series in Optical Sciences (Springer, 2011).

[54] W. Drexler and J. Fujimoto, *Optical Coherence Tomography: Technology and Applications*, Biological and medical physics, biomedical engineering (Springer, 2008).

[55] R. Bracewell, *The Fourier transform and its applications*, McGraw-Hill series in electrical and computer engineering (McGraw Hill, 2000).

[56] J. W. Cooley and J. W. Tukey, "An algorithm for the machine calculation of complex Fourier series," Math. Comp. **19**, 297–301 (1965).

[57] M. Frigo and S. G. Johnson, "The design and implementation of FFTW3," Proceedings of the IEEE **93**, 216–231 (2005). Special issue on "Program Generation, Optimization, and Platform Adaptation".

[58] D. Potts, G. Steidl, and M. Tasche, "Fast Fourier transforms for nonequispaced data: A tutorial," in *Modern Sampling Theory: Mathematics and Applications*, J. Benedetto and P. J. S. G. Ferreira, eds. (Birkhäuser, 2001), chap. 12, pp. 247–269.

[59] S. Kunis, "Nonequispaced FFT — generalisation and inversion," Ph.D. thesis, Universität Lübeck (2006).

[60] J. Keiner, S. Kunis, and D. Potts, "Using NFFT 3—a software library for various nonequispaced fast Fourier transforms," ACM Trans. Math. Softw. **36**, 19:1–19:30 (2009).

[61] G. C. Sherman, "Application of the convolution theorem to Rayleigh's integral formulas," J. Opt. Soc. Am. **57**, 546–547 (1967).

[62] J. Goodman, *Statistical optics*, Wiley series in pure and applied optics (Wiley, 1985).

[63] F. Byron and R. Fuller, *Mathematics of Classical and Quantum Physics*, Dover Books on Physics (Dover Publications, 1992).

[64] E. Wolf, "Three-dimensional structure determination of semi-transparent objects from holographic data," Opt. Commun. **1**, 153–156 (1969).

[65] W. H. Carter, "Computational reconstruction of scattering objects from holograms," J. Opt. Soc. Am. **60**, 306–314 (1970).

[66] P.-C. Ho and W. H. Carter, "Structural measurement by inverse scattering in the first Born approximation," Appl. Opt. **15**, 313–314 (1976).

[67] A. F. Fercher, H. Bartelt, H. Becker, and E. Wiltschko, "Image formation by inversion of scattered field data: experiments and computational simulation," Appl. Opt. **18**, 2427–2439 (1979).

[68] R. N. Wilke, M. Priebe, M. Bartels, K. Giewekemeyer, A. Diaz, P. Karvinen, and T. Salditt, "Hard x-ray imaging of bacterial cells: nano-diffraction and ptychographic reconstruction," Opt. Express **20**, 19232–19254 (2012).

[69] R. Mueller, M. Kaveh, and G. Wade, "Reconstructive tomography and applications to ultrasonics," Proc. IEEE **67**, 567 – 587 (1979).

[70] A. Devaney, "A filtered backpropagation algorithm for diffraction tomography," Ultrasonic Imaging **4**, 336–350 (1982).

[71] T. Lo and P. Inderwiesen, *Fundamentals of Seismic Tomography*, Geophysical Monograph Series (Society of Exploration Geophysicists, 1994).

[72] E. Adelson and J. Wang, "Single lens stereo with a plenoptic camera," Pattern Analysis and Machine Intelligence, IEEE Transactions on **14**, 99–106 (1992).

[73] R. Ng, M. Levoy, M. Brédif, G. Duval, M. Horowitz, and P. Hanrahan, "Light field photography with a hand-held plenoptic camera," Tech. rep., Stanford University (2005).

[74] D. Gábor, "A new microscopic principle," Nature **161**, 777–778 (1948).

[75] D. Gábor, "Microscopy by reconstructed wave-fronts," R. Soc. Lond. Proc. Ser. A Math. Phys. Eng. Sci. **197**, 454–487 (1949).

[76] Y. N. Denisyuk, "On the reflection of optical properties of an object in a wave field of light scattered by it," Proc. U.S.S.R. Acad. Sci. **144**, 1275–1278 (1962).

[77] E. N. Leith and J. Upatnieks, "Reconstructed wavefronts and communication theory," J. Opt. Soc. Am. **52**, 1123–1128 (1962).

[78] D. Gábor, *Holography, 1948-1971: Nobel Lecture*, Nobel lectures (Norstedt, 1972).

[79] J. W. Goodman and R. W. Lawrence, "Digital image formation from electronically detected holograms," Appl. Phys. Lett. **11**, 77–79 (1967).

[80] A. Fercher, C. Hitzenberger, G. Kamp, and S. El-Zaiat, "Measurement of intraocular distances by backscattering spectral interferometry," Opt. Commun. **117**, 43–48 (1995).

[81] S. R. Chinn, E. A. Swanson, and J. G. Fujimoto, "Optical coherence tomography using a frequency-tunable optical source," Opt. Lett. **22**, 340–342 (1997).

[82] M. A. Bail, G. Häusler, J. M. Herrmann, F. Kiesewetter, M. W. Lindner, and A. Schultz, "Optical coherence tomography by spectral radar for the analysis of human skin," Proc. SPIE **3196**, 38–49 (1998).

[83] R. Leitgeb, C. Hitzenberger, and A. Fercher, "Performance of Fourier domain vs. time domain optical coherence tomography," Opt. Express **11**, 889–894 (2003).

[84] M. Choma, M. Sarunic, C. Yang, and J. Izatt, "Sensitivity advantage of swept source and Fourier domain optical coherence tomography," Opt. Express **11**, 2183–2189 (2003).

[85] J. F. de Boer, B. Cense, B. H. Park, M. C. Pierce, G. J. Tearney, and B. E. Bouma, "Improved signal-to-noise ratio in spectral-domain compared with time-domain optical coherence tomography," Opt. Lett. **28**, 2067–2069 (2003).

[86] T. Wilson, *Confocal microscopy* (Academic Press, 1990).

[87] M. Gu, C. J. R. Sheppard, and X. Gan, "Image formation in a fiber-optical confocal scanning microscope," J. Opt. Soc. Am. A **8**, 1755–1761 (1991).

[88] M. Gu, *Principles of Three Dimensional Imaging in Confocal Microscopes* (World Scientific, 1996).

[89] A.-H. Dhalla, J. V. Migacz, and J. A. Izatt, "Crosstalk rejection in parallel optical coherence tomography using spatially incoherent illumination with partially coherent sources," Opt. Lett. **35**, 2305–2307 (2010).

[90] D. Hillmann, G. Hüttmann, and P. Koch, "Using nonequispaced fast Fourier transformation to process optical coherence tomography signals," Proc. SPIE p. 73720R (2009).

[91] Z. Hu and A. M. Rollins, "Fourier domain optical coherence tomography with a linear-in-wavenumber spectrometer," Opt. Lett. **32**, 3525–3527 (2007).

[92] J. Xi, L. Huo, J. Li, and X. Li, "Generic real-time uniform k-space sampling method for high-speed swept-source optical coherence tomography," Opt. Express **18**, 9511–9517 (2010).

[93] B. Park, M. C. Pierce, B. Cense, S.-H. Yun, M. Mujat, G. Tearney, B. Bouma, and J. de Boer, "Real-time fiber-based multi-functional spectral-domain optical coherence tomography at 1.3⁻m," Opt. Express **13**, 3931–3944 (2005).

[94] M. Mujat, B. H. Park, B. Cense, T. C. Chen, and J. F. de Boer, "Autocalibration of spectral-domain optical coherence tomography spectrometers for in vivo quantitative retinal nerve fiber layer birefringence determination," Journal of Biomedical Optics **12**, 041205–041206 (2007).

[95] S. Makita, T. Fabritius, and Y. Yasuno, "Full-range, high-speed, high-resolution 1-μm spectral-domain optical coherence tomography using BM-scan for volumetric imaging of the human posterior eye," Opt. Express **16**, 8406–8420 (2008).

[96] B. Hofer, B. Považay, B. Hermann, A. Unterhuber, G. Matz, and W. Drexler, "Dispersion encoded full range frequency domain optical coherence tomography," Opt. Express **17**, 7–24 (2009).

[97] D. Hillmann, T. Bonin, C. Lührs, G. Franke, M. Hagen-Eggert, P. Koch, and G. Hüttmann, "Common approach for compensation of axial motion artifacts in swept-source OCT and dispersion in Fourier-domain OCT," Opt. Express **20**, 6761–6776 (2012).

[98] S. H. Yun, G. Tearney, J. de Boer, and B. Bouma, "Motion artifacts in optical coherence tomography with frequency-domain ranging," Opt. Express **12**, 2977–2998 (2004).

[99] R. Yadav, K.-S. Lee, J. P. Rolland, J. M. Zavislan, J. V. Aquavella, and G. Yoon, "Micrometer axial resolution OCT for corneal imaging," Biomed. Opt. Express **2**, 3037–3046 (2011).

[100] T. Hillman and D. Sampson, "The effect of water dispersion and absorption on axial resolution in ultrahigh-resolution optical coherence tomography," Opt. Express **13**, 1860–1874 (2005).

[101] P. Puvanathasan, P. Forbes, Z. Ren, D. Malchow, S. Boyd, and K. Bizheva, "High-speed, high-resolution Fourier-domain optical coherence tomography system for retinal imaging in the 1060 nm wavelength region," Opt. Lett. **33**, 2479–2481 (2008).

[102] W. Benjamin and I. Borish, *Borish's clinical refraction* (Butterworth-Heinemann/Elsevier, 2006).

[103] A. Yang, F. Vanholsbeeck, S. Coen, and J. Schroeder, "Chromatic dispersion compensation of an OCT system with a programmable spectral filter," Proc. SPIE **8091**, 809125 (2011).

[104] M. Wojtkowski, V. Srinivasan, T. Ko, J. Fujimoto, A. Kowalczyk, and J. Duker, "Ultrahigh-resolution, high-speed, Fourier domain optical coherence tomography and methods for dispersion compensation," Opt. Express **12**, 2404–2422 (2004).

[105] Y. Yasuno, S. Makita, T. Endo, G. Aoki, M. Itoh, and T. Yatagai, "Simultaneous B-M-mode scanning method for real-time full-range Fourier domain optical coherence tomography," Appl. Opt. **45**, 1861–1865 (2006).

[106] F. J. Harris, "On the use of windows for harmonic analysis with the discrete Fourier transform," Proc. IEEE **66**, 51–83 (1978).

[107] K. Zhang and J. U. Kang, "Graphics processing unit accelerated non-uniform fast fourier transform for ultrahigh-speed, real-time Fourier-domain oct," Opt. Express **18**, 23472–23487 (2010).

[108] K. Zhang and J. U. Kang, "Real-time intraoperative 4D full-range FD-OCT based on the dual graphics processing units architecture for microsurgery guidance," Biomed. Opt. Express **2**, 764–770 (2011).

[109] D. Hillmann, C. Lührs, T. Bonin, P. Koch, and G. Hüttmann, "Holoscopy—holographic optical coherence tomography," Opt. Lett. **36**, 2390–2392 (2011).

[110] D. Hillmann, G. Franke, C. Lührs, P. Koch, and G. Hüttmann, "Efficient holoscopy image reconstruction," Opt. Express **20**, 21247–21263 (2012).

[111] T. van Leeuwen, D. Faber, and M. Aalders, "Measurement of the axial point spread function in scattering media using single-mode fiber-based optical coherence tomography," IEEE J. Sel. Top. Quant. Eletron. **9**, 227–233 (2003).

[112] P. Ferraro, G. Coppola, S. D. Nicola, A. Finizio, and G. Pierattini, "Digital holographic microscope with automatic focus tracking by detecting sample displacement in real time," Opt. Lett. **28**, 1257–1259 (2003).

[113] T. Colomb, N. Pavillon, J. Kühn, E. Cuche, C. Depeursinge, and Y. Emery, "Extended depth-of-focus by digital holographic microscopy," Opt. Lett. **35**, 1840–1842 (2010).

[114] P. Langehanenberg, G. Bally, and B. Kemper, "Autofocusing in digital holographic microscopy," 3D Res. **2**, 27:1–27:11 (2011).

[115] C. Lührs, "Holoscopy – holographic optical coherence tomography," Diploma thesis, Fachhochschule Lübeck (2011).

[116] P. D. Woolliams, R. A. Ferguson, C. Hart, A. Grimwood, and P. H. Tomlins, "Spatially deconvolved optical coherence tomography," Appl. Opt. **49**, 2014–2021 (2010).

[117] E. Botcherby, R. Juškaitis, M. Booth, and T. Wilson, "Aberration free refocusing for high numerical aperture microscopy," Proc. SPIE **6861**, 686110–686112 (2008).

[118] A. Stadelmaier and J. H. Massig, "Compensation of lens aberrations in digital holography," Opt. Lett. **25**, 1630–1632 (2000).

[119] T. Colomb, F. Montfort, J. Kühn, N. Aspert, E. Cuche, A. Marian, F. Charrière, S. Bourquin, P. Marquet, and C. Depeursinge, "Numerical parametric lens for shifting, magnification, and complete aberration compensation in digital holographic microscopy," J. Opt. Soc. Am. A **23**, 3177–3190 (2006).

[120] T. Colomb, J. Kühn, F. Charrière, C. Depeursinge, P. Marquet, and N. Aspert, "Total aberrations compensation in digital holographic microscopy with a reference conjugated hologram," Opt. Express **14**, 4300–4306 (2006).

[121] S. Weinberg, *Lectures on Quantum Mechanics* (Cambridge University Press, 2012).

[122] W. H. Press, S. A. Teukolsky, W. T. Vetterling, and B. P. Flannery, *Numerical Recipes 3rd Edition: The Art of Scientific Computing* (Cambridge University Press, New York, NY, USA, 2007), 3rd ed.

Publications

First author

- D. Hillmann, G. Hüttmann, and P. Koch, "Using nonequispaced fast Fourier transformation to process optical coherence tomography signals," Proc. SPIE **7372**, 73720R (2009).

- D. Hillmann, C. Lührs, T. Bonin, P. Koch, A. Vogel, and G. Hüttmann, "Holoscopy: holographic optical coherence tomography," Proc. SPIE **8091**, 80911H (2011).

- D. Hillmann, C. Lührs, T. Bonin, P. Koch, and G. Hüttmann, "Holoscopy—holographic optical coherence tomography," Opt. Lett. **36**, 2390–2392 (2011).

- D. Hillmann, T. Bonin, C. Lührs, G. Franke, M. Hagen-Eggert, P. Koch, and G. Hüttmann, "Common approach for compensation of axial motion artifacts in swept-source OCT and dispersion in Fourier-domain OCT," Opt. Express **20**, 6761–6776 (2012).

- D. Hillmann, G. Franke, C. Lührs, P. Koch, and G. Hüttmann, "Efficient holoscopy image reconstruction," Opt. Express **20**, 21247–21263 (2012).

Co-author

- G. Hüttmann, G. Franke, C. Lührs, P. Koch, and D. Hillmann, "3D optical imaging: Holoscopy makes ultrafast lensless imaging of scattering tissues possible," Laser Focus World **48** (2012).

- G. Hüttmann, J. Probst, T. Just, H. Pau, S. Oelckers, D. Hillmann, P. Koch, and E. Lankenau, "Real-time volumetric optical coherence tomography OCT imaging with a surgical microscope," Head Neck Oncol. **2**, O8 (2010).

- J. Probst, D. Hillmann, E. Lankenau, C. Winter, S. Oelckers, P. Koch, and G. Hüttmann, "Optical coherence tomography with online visualization of more than seven rendered volumes per second," J. Biomed. Opt. **15**, 026014 (2010).

- G. L. Franke, D. Hillmann, T. Claußen, C. Lührs, P. Koch, and G. Hüttmann, "High resolution holoscopy," Proc. SPIE **8213**, 821324 (2012).

- M. Hagen-Eggert, D. Hillmann, P. Koch, and G. Hüttmann, "Diffusion-sensitive Fourier-domain optical coherence tomography," Proc. SPIE **7889**, 78892B (2011).

Talks

- D. Hillmann, G. Hüttmann, and P. Koch, "Using nonequispaced fast Fourier transformation to process optical coherence tomography signals," *European Conference on Biomedical Optics ECBO 2009*, Jun 14–18, 2009, Munich, Germany

- D. Hillmann, C. Lührs, T. Bonin, P. Koch, A. Vogel, and G. Hüttmann, "Holoscopy: holographic optical coherence tomography," *European Conference on Biomedical Optics ECBO 2011*, May 22–26, 2011, Munich, Germany

- D. Hillmann, T. Bonin, G. Franke, M. Hagen-Eggert, P. Koch, and G. Hüttmann, "Numerical movement correction for swept-source full-field optical coherence tomography," *SPIE BiOS, part of SPIE Photonics West 2011*, Jan 22–27, 2011, San Francisco, USA

- D. Hillmann, G. Franke, C. Lührs, T. Claußen, P. Koch, and G. Hüttmann, "Holoscopy image reconstruction," *SPIE BiOS, part of SPIE Photonics West 2012*, Jan 21–26, 2012, San Francisco, USA

Patents

- D. Hillmann, G. Hüttmann, P. Koch, C. Lührs, and A. Vogel, "Verfahren zur optischen Tomographie," Deutsche Patentschrift DE 10 2011 018 0603 B3 2012.06.06, applied Apr 21, 2011.

Danksagung

Zahlreiche Menschen haben mir bei dieser Arbeit direkt oder indirekt geholfen. An dieser Stelle möchte ich mich bei ihnen bedanken und die Wichtigsten benennen.

Mein ganz besonderer Dank gebührt dabei meinem Betreuer während der Doktorarbeit, Gereon Hüttmann. Seine zahlreichen Ideen und Visionen haben diese Arbeit erst möglich gemacht. Zudem hat er mir immer mit Rat zur Seite gestanden und konnte bei vielen Problemen helfen.

Der Thorlabs GmbH, bzw. der damaligen Thorlabs HL AG, danke ich, dass sie mich bei dieser Arbeit unterstützt hat. Besonderer Dank geht an Peter Koch und Jörn Wollenzin. Von Peter habe ich viel über OCT gelernt, und er hat mich dazu gebracht, mit dieser Arbeit anzufangen. Beide, Peter und Jörn, haben mich bei der Arbeit bestmöglich unterstützt.

Den ehemaligen und aktuellen Leitern des Instituts für Biomedizinische Optik in Lübeck und des Medizinischen Laserzentrums Lübeck, Reginald Birngruber, Alfred Vogel und Ralf Brinkmann, danke ich dafür, dass ich die Arbeit am BMO bzw. MLL durchführen durfte. Außerdem möchte ich mich natürlich beim Vorsitzenden des Prüfungsausschusses Alfred Mertins und bei den Berichterstattern Alfred Vogel, Thorsten Buzug und Rainer Heintzmann bedanken, dass sie sich zu diesen Aufgaben bereit erklärt haben.

Für die hervorragende Vorarbeit im Bereich Full-Field OCT danke ich Tim Bonin, Martin Hagen-Eggert und Gesa Franke. Auch bei den Kollegen, die an der Holoskopie mitgewirkt haben, möchte ich mich besonders bedanken, insbesondere bei Christian Lührs und Gesa Franke. Christian war an vielen Messungen beteiligt und hat den losen Laboraufbau in einen messetauglichen Prototypen verwandelt. Gesa hat mit an der hochauflösenden Holoskopie gearbeitet und versucht, diese nun weiter zu bringen.

Zahlreiche weitere Kollegen, sowohl bei Thorlabs, als auch beim BMO/MLL, haben mir mit wertvollen Diskussionen, Tipps, Rat und Tat geholfen. An dieser Stelle möchte ich noch Christian Winter nennen, von dem ich einiges über Optik und OCT gelernt habe.

Nicht zuletzt möchte ich mich bei meinen Eltern bedanken, die mich stets ermutigt, meine Interessen gefördert und mich bei allem unterstützt haben. Sie haben mir mein Studium und damit diese Arbeit überhaupt erst ermöglicht.

Aktuelle Forschung Medizintechnik

Herausgeber:

Prof. Dr. Thorsten M. Buzug

Institut für Medizintechnik, Universität zu Lübeck

Themen

Werke aus folgenden Themengebieten werden gerne in die Reihe aufgenommen: Biomedizinische Mikro- und Nanosysteme, Elektromedizin, biomedizinische Mess- und Sensortechnik, Monitoring, Lasertechnik, Robotik, minimalinvasive Chirurgie, integrierte OP-Systeme, bildgebende Verfahren, digitale Bildverarbeitung und Visualisierung, Kommunikations- und Informationssysteme, Telemedizin, eHealth und wissensbasierte Systeme, Biosignalverarbeitung, Modellierung und Simulation, Biomechanik, aktive und passive Implantate, Tissue Engineering, Neuroprothetik, Dosimetrie, Strahlenschutz, Strahlentherapie.

Autorinnen und Autoren

Autoren der Reihe sind in der Regel junge Promovierte und Habilitierte, die exzellente Abschlussarbeiten verfasst haben.

Leserschaft

Die Reihe wendet sich einerseits an Studierende, Promovenden und Habilitanden aus den Bereichen Medizintechnik, Medizinische Ingenieurwissenschaft, Medizinische Physik, Medizinische Informatik oder ähnlicher Richtungen. Andererseits stellt die Reihe aktuelle Arbeiten aus einem sich schnell entwickelnden Feld dar, so dass auch Wissenschaftlerinnen und Wissenschaftler sowie Entwicklerinnen und Entwickler an Universitäten, in außeruniversitären Forschungseinrichtungen und der Industrie von den ausgewählten Arbeiten in innovativen Gebieten der Medizintechnik profitieren werden.

Begutachtungsprozess

Die Qualitätssicherung erfolgt in drei Schritten. Zunächst werden nur Arbeiten angenommen die mindestens magna cum laude bewertet sind. Im zweiten Schritt wird ein Mitglied des Editorial Boards die Annahme oder Ablehnung des Werkes empfehlen. Im letzten Schritt wird der Reihenherausgeber über die Annahme oder Ablehnung entscheiden sowie Änderungen in der Druckfassung empfehlen. Die Koordination übernimmt der Reihenherausgeber.

Kontakt

Prof. Dr. Thorsten M. Buzug
Institut für Medizintechnik
Universität zu Lübeck
Ratzeburger Allee 160
23538 Lübeck, Germany

Tel.:   +49 (0) 451 / 500-5400
Fax: +49 (0) 451 / 500-5403
E-Mail: buzug@imt.uni-luebeck.de
Web: http://www.imt.uni-luebeck.de

Stand: Januar 2014. Änderungen vorbehalten.
Erhältlich im Buchhandel oder beim Verlag.

Abraham-Lincoln-Straße 46
D-65189 Wiesbaden
Tel. +49 (0)6221. 345 - 4301
www.springer-vieweg.de